T0418652

DREAMS OF PRESENCE

Dreams of Presence

A Geographical Theory of Culture

MITCH ROSE

UNIVERSITY OF TORONTO PRESS
Toronto Buffalo London

Toronto Buffalo London
utorontopress.com
Printed in the USA

ISBN 978-1-4875-6617-3 (cloth)
ISBN 978-1-4875-6967-9 (EPUB)
ISBN 978-1-4875-6937-2 (PDF)

Library and Archives Canada Cataloguing in Publication

Title: Dreams of presence : a geographical theory of culture / Mitch Rose.
Names: Rose, Mitch, author
Description: Includes bibliographical references and index.
Identifiers: Canadiana (print) 20240471830 | Canadiana (ebook) 20240471911 | ISBN 9781487566173 (hardcover) | ISBN 9781487569679 (EPUB) | ISBN 9781487569372 (PDF)
Subjects: LCSH: Cultural geography. | LCSH: Culture.
Classification: LCC GF50 .R67 2024 | DDC 304.2–dc23

Cover design: Will Brown
Cover image: National Maritime Museum, Greenwich, London

We wish to acknowledge the land on which the University of Toronto Press operates. This land is the traditional territory of the Wendat, the Anishnaabeg, the Haudenosaunee, the Métis, and the Mississaugas of the Credit First Nation.

University of Toronto Press acknowledges the financial support of the Government of Canada, the Canada Council for the Arts, and the Ontario Arts Council, an agency of the Government of Ontario, for its publishing activities.

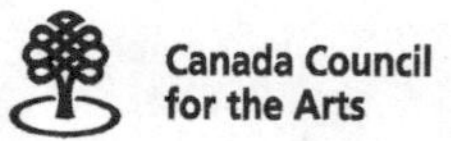

For Liz, Toby, and Mimi

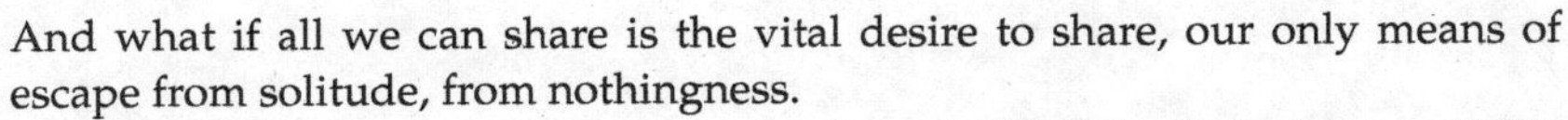

And what if all we can share is the vital desire to share, our only means of escape from solitude, from nothingness.

– Edmond Jabès, *The Book of Shares* (1989)

If the notion of solidarity is an empirical one, a notion which the sociologist has the right to deal with, it is also very mysterious.

– Emmanuel Levinas, "Transcendence and Height" (1996b)

Contents

Preface

This project began in 2006. About eighteen years ago. Something about this statement makes me feel obliged to poeticize about the value of slow thinking. But I am not sure I want to make a virtue of my long history of feints and deferrals. Anyone that knows me (colleagues and friends) knows I have been talking about this book for a long time. It began as an Arts and Humanities Research Council fellowship where I proposed to apply my ethnographic work in Egypt to a novel approach to landscape. Almost immediately, however, I realized that the concept of landscape I wanted to develop could not be contained within a chapter. This is because (I also realized) the theory I wanted to develop was not a theory of landscape but was a theory of culture. And attempting to theorize culture (my final realization) could only be a project of catastrophic folly. Culture, after all, was a concept in terminal decline. Critiqued for its essentialism, its reductivism, its humanism, and its cognitive bias, it was wholly out of step with the process ontologies currently in ascendency; frameworks that emphasize transformation, movement, and becoming, not the appeals to organicism, history, and stasis that are necessarily hung around the culture concept. Thus, while the question of culture beckoned, answering it did not seem possible at the time. To be clear, I am not ready to defend the almost twenty-year period it took me to write this work. But I can allow myself to acknowledge that there were a lot of pieces to put together. And that some things just take a while.

There are two things, I think, that make the concept of culture developed here distinctive. The first is that it is existential in nature, and the second is that it is fundamentally reliant on geography. In terms of the former, understanding what I mean by an existential approach takes up most of the space in the book. While the approach is not wholly novel – the anthropologist Michael Jackson's work in existentialist anthropology has many cross-overs – it does cut against the dominance of process ontologies currently defining the social science landscape. By attending to the existential limits of being a living being, this project approaches culture as a response to a set of universal problems that are conceptualized as both non-relational and non-negotiable. In this sense, this book continues along a path that I (and others) have been developing for a while now on *the negative* (see Bissell, 2009, 2022; Dekeyser & Jellis, 2021; Dekeyser et al., 2022; Harrison, 2007a, 2015; Joronen & Rose, 2021; Philo, 2017; Pugh & Chandler, 2023; Rose, 2014; Rose et al., 2021; Wylie, 2009). Culture, I argue here, is a response to the problems posed by these limits. It is how we cope with the wound of not being self-standing sovereign beings. In terms of geography, one of the abiding themes of the text is that culture is not an interiorized identity that is secondarily expressed through external forms. On the contrary, culture is something built. It is something that takes place through external forms because those forms give culture a visible, material reality. Culture, in this sense, is akin to an icon: a material structure that gives phenomenality to a non-material idea. While the terminology of "icon" will be familiar to many cultural geographers, it is a position that is arrived at via very different terms than those used by Cosgrove (1983, 1985) and Daniels (1989). And yet, even with that being said, the framework does not radically alter our conception of what cultural *is* within our traditional geographic frames. On the contrary, it reinforces what cultural geographers have known for a very long time, specifically, that geography is fundamental to culture – even if it explains that fundamental relation in very different terms. In this sense, my hope is that this book will reopen the question of culture both in geography and in the social sciences more broadly. At a minimum, the book endeavours to illustrate why the question of culture abides. While I agree that the culture concept has a long and problematic history (which I discuss), the problem that culture poses (the problem of our difference and why human beings care for, defend, and identify with those differences) is one that has not gone away. The ambition of this book (and the twenty-year project it represents) is not simply to provide an answer to this problem but to illustrate how geography is key – indeed fundamental – to its appraisal.

Before moving on to the acknowledgments, I would like to provide a quick roadmap through the chapters. While I would love for everyone to be so seduced by this book's arguments and felicitous prose that they voraciously read it cover to cover, I am fully aware of academic pressures. Indeed, while Derrida chastizes the "bad reader," I would say better to be read badly than not read at all. Thus, rather than have readers guessing about where to begin, I offer the following guidance.

The overall structure of the book is three sections topped and tailed by an introduction and conclusion. Each section represents a key manoeuvre in the argument. Section 1 discusses how the concept of culture has been developed in anthropology, geography, and other contemporary literatures and illustrates how and why those approaches need to change. Section 2 develops the philosophical architecture of the argument, and section 3 distils that architecture into a coherent theory of culture. While the introduction provides a summary of each chapter, here I give some guidance on what I think are the pivotal moments in the argument so you can plan your reading.

The aim of section 1 is to illustrate how the question of culture needs to move from a question of difference to a question of claiming. To some extent, what I mean by this shift is explained in chapter 1, so that may be enough to understand the central point of the section. The structure of each chapter is roughly the same; they each endeavour to illustrate (1) how the question of difference is a dead-end and (2) why the question of claiming is a more productive angle. The bulk of each chapter is devoted to the former. This is because illustrating how the question of difference evolves in each tradition requires some excavation. If, however, you are primarily interested in what I mean by claiming (and why it is a more productive angle), the final section of each chapter may be good enough. You may also get the gist of what I am doing from one of the chapters and thus not feel compelled to read all three. If you were going to choose one, I might choose chapter 4 only because I think the distinctive trajectory of this work is developed most fully there. The aim of section 2 is to introduce a distinctive theory of the subject. The material that is most relevant here can be found in chapter 6. This chapter provides a distinctive interpretation of the dwelling concept as it arises in the work of Emmanuel Levinas; a framework I will use to develop the theory of culture in chapters 7 and 8. Chapter 5 explains the origin of the dwelling concept and how it evolves through the work of Martin Heidegger. Can you understand chapter 6 without chapter 5? I think so, particularly if you know something about Heidegger. That said, I am putting forth my own interpretation of the dwelling concept as it evolves in his work so some nuance may be missed. The final section (section 3) is the money section and both chapters (7 and 8) are key to the argument. While the former explains how I understand culture, the latter explains the role of geography. Is it possible to skip from chapter 1 to chapters 7–8? It's possible but the existential demand that underpins the argument will be underappreciated (that requires chapter 6). Finally, chapter 9 is written as a pithy summary of the argument. At this point I am no longer defending the central pillars of the thesis but discussing its distinctive features and implications. If you want to know what this book is essentially about (without worry about how I defend it), then chapters 1 and 9 could potentially do it, though I think chapters 1, 6, 7, and 8 would be most useful as they constitute the beating heart of the argument.

In terms of recognition and gratitude, it is difficult to acknowledge all the helping hands and forms of encouragement that have boosted this project along the way. Though if I were to pinpoint the force that ultimately instigated me to finish this project, it would be the inspiring group of colleagues-friends I have had the pleasure of working with over the decade and many much longer. Writing in earnest began after visiting fellowships by David Bissell and Mikko Joronen at Aberystwyth University in 2018. This was a highly productive time that led to numerous conversations and collaborations, mostly notably the publication of *Negative Geographies: Exploring the Politics of Limits* (2021) that David and I coedited with Paul Harrison. Indeed, the imprint of Paul, David, and Mikko can be found all over the key arguments that underpin this project and I am extremely grateful for their friendship and contributions over the years. Other important interlocutors are John Wylie, Chris Philo, Matt Hannah, Jessica Dubow, Ben Anderson, Anna Secor, Thomas Dekeyser, and Vickie Zhang. One of my reviewers suggested that this book ties together many of the papers and chapters I have written over the years. If this is true (and I think it is) then the collaborations and conversations I have had with these individuals have played an important part in developing its various pieces.

Other important contributors are colleagues in the Department of Geography and Earth Sciences at Aberystwyth University, particularly Pete Merriman, Mark Whitehead, Mike Woods, Gareth Hoskins, Jesse Heley, Rhys A. Jones, and Rhys D. Jones. I also want to acknowledge the invaluable insight I received from readers who were generous enough to comment on draft chapters. These include Elizabeth Gagen and Mark Whitehead (chapter 1), Jenny Pickerill (chapter 2), Pete Merriman (chapter 4), Mikko Joronen (chapter 5), Paul Harrison (chapter 6), David Bissell (chapter 7), and Dave Clarke (chapter 8). I would also like to thank my editor Jodi Lewchuk at the University of Toronto Press for being so supportive of this book from the beginning (and for trying so hard to make it open access) as well as *Progress in Human Geography* and *Environment and Planning D: Society and Space* where versions of chapters 3 (PGH), 4 and 5 (EPD) were initially published. A quick shout-out to Jimmy Roth who pointed me to the work of the poet Edmond Jabès (see epigram) and a long overdue acknowledgment to the AHRC who many (many) years ago provided the support that initiated this whole endeavour.

Finally, there are two people (that I know of) who have taken the time to read the manuscript in its entirety. The first is Chris Philo whose generous and incisive comments were instrumental not only in helping me refine the argument (particularly around issues of disciplinary history) but also for giving me confidence that its various threads hung together. The second is Edith Franklin whose careful reading and reflective editing helped ensure the monograph would be accessible to a wide audience. My biggest debt, however, goes to Elizabeth Gagen. It is difficult to put into words her input into this book. While she has certainly read many drafts, commented on early chapters, and commented on rewrites of those chapters, she has also been there, from the beginning, telling me it would get done. I have never been someone whose work and life easily separate (or balance). But within the normal storm of living and working together, Liz has always been a thoughtful, sensitive, and judicious guide. And for that (and many things), I am eternally grateful.

DREAMS OF PRESENCE

Chapter One

Culture War, Culture Loss

No matter how much one trains one's attention on the supposedly hard facts of social existence … the supposedly soft facts of that existence, what do people imagine human life to be all about, how do they think people ought to live, what grounds beliefs, legitimizes punishment, sustains hope, or accounts for loss, crowd in … One can ignore such facts, obscure them or pronounce them forceless. But they do not thereby go away. Whatever the infirmities of the concept of culture … there is nothing for it but to persist in spite of them. Tone deafness, willed or congenital, and however belligerent, will not do.

– Clifford Geertz, *After the Fact* (1995)

A Lament

In a 2008 interview in *Cultural Anthropology*, professor George Marcus (of Clifford and Marcus fame) argued that the discipline of anthropology, the discipline that stands as the traditional heartland for theorizing the concept of culture, is in what he termed theoretical *suspension*: "there are no new ideas and none on the horizon, as well as no indication that [anthropology's] traditional stock of knowledge shows any sign of revitalization … even the concept of culture, while emblematic of what the discipline is interested in, is no longer viable analytically and has been appropriated everywhere and by everybody" (Marcus, 2008, p. 3). One has to be careful of making too much of such statements, of concerns voiced in the informality of an interview rather than the tempered space of a single-authored article. The quote represents the view of one professor in an important corner of an international discipline, a discipline that, as Marcus states, has also shown an incredible amount of invention. To suggest that anthropology is in theoretical suspension belies the influence that the discipline has lent to both geography and the social sciences more broadly, particularly the work of Tim Ingold (2000, 2011), Daniel Miller (2005, 2008), Kathleen Stewart (1996, 2007), Anna Tsing (2015a), and, more recently, Eduardo Vivieros de Castro (2014), Povinelli (2002), and many others (e.g., Bender & Aitken, 1998; Taussig, 2004; Tilley, 2004). But the claim that the concept of culture is no longer analytically viable is less easy to dismiss. Whatever we may think about Marcus's lament, indeed whether we think that the loss of culture as a viable concept is lamentable, there is something resonant about his claim. With all the theoretical developments that have happened in anthropology, cultural geography, and Cultural Studies over the last decade, one cannot deny that the inclination to theoretically define and/or explain *culture*, an inclination that has been a central pillar of social scientific thought for much of the previous century, has, if not waned, then certainly become an increasingly oblique concern.

To some extent, culture's demise can be laid at the feet of anthropology, the traditional keeper and purveyor of cultural theory, which turned its back on the concept in the 1990s; a somewhat discrete disciplinary event that nonetheless had much wider reverberations (M.S. Archer, 1996; Behar & Gordon, 1995; Minkov & Hofstede, 2012; Richardson, 1988; Sewell, 2005; P.B. Smith, 2004; Spiegel, 2005). It can also be attributed to a broader ontological turn happening throughout the social sciences, a movement represented by non-representational theory in geography (Anderson & Harrison, 2010; Dewsbury et al., 2002; Escobar, 2007; Marston et al., 2005; Thrift, 2008), the Indigenous ontologies literature in anthropology (Alberti et al., 2011; Henare et al., 2007; Holbraad & Pedersen, 2017; Sivado, 2015; Viveiros de Castro, 2014), and a broader turn towards ontological arguments in feminist studies (Åsberg, 2010; Colebrook, 2000; Hemmings, 2005; Lawson, 2003; M. McNeil, 2010; Pedwell & Whitehead, 2012; Witz, 2000), sociology (Feely, 2016; Kim, 2019; Marres, 2009; Pickering, 2017; Savransky, 2017) and science studies (Sismondo, 2015; Van Heur et al., 2013; Woolgar & Lezaun, 2015). To be sure, there is a history to be written about the culture concept's rise and fall. But at the heart of this history (and perhaps at the heart of any such history) will be the simple fact that the term came to be increasingly incoherent as a formal analytic tool, simultaneously being used to explain everything while watering down the organizing mechanics that could be used to explain anything. As the geographer Don Mitchell (1995) famously argued, the culture concept had become an essentially ungrounded term, managing to hold social relations together while providing no account of the generative forces that facilitate the holding: "culture is seen as a level, medium or idiom but nowhere ... is there a theoretical discussion of what constitutes these spheres" (p. 100). Sage's most recent edition of the *Handbook of Cultural Geography* (Anderson et al., 2003) (itself almost twenty years old) seems written to reinforce Mitchell's point. Here, culture is defined as an "an unfolding intellectual terrain" that moots a number of key words (e.g., power, meaning, doing, a way of life, and a distribution of things), all of which circle around the culture concept like a black hole, unable to define the phenomenon but also unable to leave its gravitational force.

While I talk about the demise of the culture concept further in later chapters, the point I want to make here is that, although the concept has ceded this intellectual territory to other debates and themes, the problem that the culture concept was devised to address has not gone away. Indeed, it could not be more urgent. Do we not hear the war drums of culture everywhere? From the polarization of American politics to the rise of populist dictatorships, to the emergence of Brexit and nativism, not to mention the continual debates about cultural appropriation, cancel culture, and culture wars in the popular press,[1] the terminology of culture is in high circulation. And while it may be tempting to dismiss such chatter as the work of pseudo-intellectualism misusing antiquated terminology to address our current political moment, such protestations would be blind to what seem like powerful social currents. In short, there is a phenomenon here. And while the terminology being used to define what that phenomenon is may be wrong, misguided, inappropriate, and essentialist, it is not like the academy is offering anything better.

A brief demonstration of the consequence of this silence can be seen in Whitehead and Perry's (2020) and Whitehead et al.'s (2018) work on the coalition of former president Donald Trump in 2016. Here, the authors make a compelling and well-evidenced argument that the trends constituting the Trump coalition cannot be explained in strictly political terms. At the centre of their argument is that when it comes to actual issues there is remarkable elasticity among Trump voters. On abortion, civil rights, and even immigration one does not find in this coalition a coherent political agenda. Rather, what one finds is an enduring consistency in terms of values, belief systems, and investment in certain national mythologies. Their term for this belief system is Christian nationalism, an outlook that fuses "American civic life with a particular type of Christian identity" (Whitehead & Perry, 2020, p. ix).[2] This group, they argue, is not particularly orthodox or exceptionally pious but is defined by a specific eschatology that gives America a privileged place in God's eternal plan; a position which explains (and justifies) American exceptionalism, as well as its wealth and global reach. It also understands itself as a community under threat; persecuted on the world stage by radical Islam and at home by immigration, secularism, and religious pluralism. In contrast, the progressive left, while also a community mobilized by values and beliefs, does not have the same level of coherence in terms of world view nor is it a coalition guided by any enduring mythology. Because left-leaning constituencies are a broader tent, with far greater ethnic and racial diversity, it is defined more by political expediency than a normative outlook. On the contrary, the Trump coalition, Whitehead and Perry argue, can best be understood by looking at it through "a cultural frame."

Whatever one might think about Christian nationalism, its prevalence as a political force and / or its coherence as a sociological phenomenon, their study is one example of a slew of recent work examining the role of culture in fomenting practices, events and decisions that are at odds with other "harder" predictors of social behaviour – that is, education, background, income. Whether we are talking about white culture (McGhee, 2021), American culture (Sandel, 2020), capitalist culture (Suzman, 2020), "call out" culture (Banet-Weiser & Miltner, 2016; Vemuri, 2018), there seems to be evidence that group identities matter. What is missing from this discussion, however, is any in-depth understanding (or coherent theorisation) of the source of these effects, that is, a coherent conceptualization of what is meant by the term *culture*. While Whitehead and Perry (2020) have much to say about the history and dynamics of the Christian nationalist movement, as well as its values, norms, cosmology and eschatology, and they have a number of data points to illustrate how these dynamics galvanize the movement, they do not have much to say about what makes these data points *cultural*. My point here, to be clear, is not to defend their analysis. It is not to argue whether there exists a Christian nationalist culture or whether the Trump coalition can be defined as a cultural problem. My concern is more epistemological, namely, how can Whitehead and Perry make such claims when there is a lack of clear theorization about their central term? The failure, in other words, is a failure in the literature; a failure indicative of the abandonment in social science of the culture concept; an abdication that ultimately incapacitates scholars from explaining, in robust conceptual terms, verifiable empirical phenomena.

The disjuncture is indicative of what I take to be a conceptual cross-roads in contemporary social science. While the endeavour to contemplate, theorize, write about or in any way study culture in a programmatic manner has dissipated from the disciplinary landscape, the phenomenon itself remains. The question of culture – the question about how and why communities cohere, cultivate particular norms, perpetuate certain cosmologies, or narrate themselves in trans historical terms – continues to appear. There are claims to identities and cultures all around us, articulations of fidelity to immemorial beliefs, racial categories, mythic histories, and teleological destinies. Such claims force themselves into the social and political questions of our time (Bloch, 2005; Trouillot, 2003); a phenomenon many social scientists are willing to acknowledge as "there," even if they are happy to be vague and indistinct about what that "there" is (Fox & King, 2002; Knauft, 2006). Thus, even if we find culture to be a weak variable in the explanation of difference, we cannot deny the term's continued deployment in pre- or post-facto justifications for various modes of behaviour and identity (Trouillot, 2003; R. Wagner, 1981). To do so would be to equate a failure of theoretical imagination to an absence of phenomenality. As Bloch (2005) suggests (in an admonishment of contemporary anthropology), the failure of cultural theory to properly address the question of culture has transformed into a silence: "in the process of criticising the answers given, [anthropologists] have dismissed the questions and … even though they may have done this to their own private satisfaction, everybody else is still asking" (p. 10). There are many good reasons for rejecting the culture question, but to dismiss the various claims to culture we hear with an argument about such claims being unreal or misguided is not a sufficient response. It is to answer a theoretical obstacle with an empirical deafness. A tone deafness, as Geertz (1995) calls it, that will not do.

A Theory of Culture

The project laid out in this book is palliative in that it endeavours to reconfigure the question of culture in a manner that once again makes culture a problem for theory. While this is the simplest way to describe this project, it is also somewhat ridiculous to put down on paper. Indeed, there are many potential absurdities inherent to this ambition, but I will begin with two. First, there is the problem that I am a cultural geographer. I offer this as a problem not because I am ill-equipped – cultural geographers have been engaging with the culture concept for over a century – but because one cannot help feeling a bit like an imposter. Theorizing culture is, after all, anthropology's birth right; a canon founded by Boas and his ostensible rejection of geography for its so-called obsession with environment (R.C. Powell, 2015). The fact that anthropologists have mostly disavowed the culture concept does not negate the sense of recklessness. Even if it is true that we all like the idea of being interdisciplinary, it is also true that no one likes an interloper. But the issue goes beyond academic turf. Geography, as a field, does not have an obvious and discrete tradition of theorizing culture. While I excavate a geographical tradition of cultural thought in chapter 3, this is a positive exercise in recanonization. Even as geographers can point to a rich tradition of cultural theory *in* geography, and an equally long legacy of theorizing cultural *objects* (monuments, landscapes, urban designs), it does not have an obvious and readily identifiable lineage of cultural *concepts*. Thus, despite the fact that geographers have thought about culture and thought about it robustly, their theories of culture are more dispersed and speculative than we find in other disciplines.

The second reason this project feels ridiculous is its timing. If the issue above concerns a theory of culture being overdue (in geography at least), one could equally argue that it comes far too late. With the advent of relational ontologies in both geography and anthropology, it would seem that talk of different cultures – of different groups that embed distinctive boundaried identities – is and should be gone. The moment for theorizing culture has passed. And for many, it was a moment that hung on for far too long (Abu-Lughod, 1991; Carrithers et al., 2010; Kuper, 1999). This is not just the case in anthropology and Cultural Studies (see T. Bennett, 2015; Blaser, 2009; Grossberg, 2006), but also in geography. Since the rise of non-representational theory (Anderson & Harrison, 2010; Dewsbury et al. 2002; Thrift, 2008; Thrift & Dewsbury, 2000) and (more recently) new materialist geographies (Daya, 2019; Forman, 2021; Gandy & Jasper, 2017; Jamieson, 2021; Roberts, 2012), cultural geographers have become increasingly focused on topics such as affect, atmospheres, and human-non-human relations, all of which have pushed the question of culture, not simply to the margins but out of the frame completely (B. Anderson, 2016). As Wylie (2016b) suggests, "the question of 'culture' seems to have fallen into abeyance" in cultural geography, "unless I have overlooked things drastically, not many cultural geographers have grappled recently with their putative *ur*-concept" (p. 376). In this sense, it seems positively out of touch to be reviving a tradition that has been forsaken not just by the social sciences in general but by one's own tribe. The culture concept, it would seem, is dead, if not murdered, dismembered, and buried in some distant swamp of intellectual penury.

The opening proposition of this book is that the question of culture can be redeemed. But it needs to be reframed. While traditionally the question of culture had been thought of in terms of the differences people *have*, the aim of this project is to think through the differences people *claim*. The project as a whole is predicated upon this distinction. In terms of the former, culture is approached as an interiority, a set of internal and/or internalized structures, inclinations, norms or enactments that constitute a subject's difference. The second question, however, conceptualizes culture as something outside the subject, an imagination of identity that is perpetually out of reach, beyond the horizon, something sought but never found. To be clear, I am not arguing that people do not have differences. On the contrary, I would argue that people have profound differences. Bodies and minds pick-up habits – distinctive rhythms and ways of doing – that provide well-trodden solutions towards various life problems. And these solutions illuminate obvious and abiding differences. But these differences are not knowable in any definitive sense. They defy understanding and determination (by observer and native alike). This means that the differences my body expresses are not *mine*. The way I think, the habits my body inflects, the practices I unreflectively enact, these are things over which I have no say. As I will discuss in later chapters, the habits and rhythms of our bodies are elusive and inscrutable. They do not belong to me and I have no command over their inheritance or destiny.

And yet, subjects make claims about these ways of doing and thinking all the time. Subjects speak about self-hood, identity, modes of existence, and ideas about the world *as if* those things were theirs. In this sense, this project sets sail from the proposition that the habits subjects embed and the differences they express are non-correspondent to the identities that subjects claim; there exists a gap between the way our bodies and minds have learned to comport to the world and the subject's sense of sovereignty over those forms and relations. The failure of cultural theory is predicated upon the presumption that subjects and identity correspond: the idea that subjects speak for a culture that is theirs and that their speech testifies to their distinctive world. The project laid out here separates this relation. By presuming the question of difference to be unanswerable, it focuses on the question of claiming: not how people are different but why people claim those differences. If we can understand this second question, then we have a potential framework for thinking culture outside its traditional traps of essentialism and the self-present subject.

The central argument of this book, therefore, is that culture is a claim. What this means and why geography is essential to its constitution, will be mapped out in the coming chapters. But in order to signal the overall orientation of the argument, we can start with the following quote by the anthropologist Michael Jackson (2005): "Societies differ in the ways in which they manage *universally identical existential issues* – keeping body and soul alive, bringing new life into the world, coping with separation and loss, creating ontological security" (p. 379, my emphasis). As an opening gambit, I would suggest that existential issues have not been recognized enough in the social sciences and certainly not by cultural theorists. These are problems that appear in our most quotidian, unexceptional, and familiar circumstances (Rose, 2014): in the problem of other people (whose habits, needs, and desires we can never fully understand, accommodate, or deny), the problems posed by the natural world (whose dynamics are eruptive and cycles unreliable), the problems of our bodies (that require constant care and fail us nonetheless), and the problem of death (whose arrival is always surprising even when it is expected). Each of these situations illuminate a territory over which human beings have no sovereignty or dominion; a situation that is both pre-ontological (because it is both anterior and exterior to all human knowledge) and non-relational (because it infinitely transcends human capacities) (see Harrison, 2007a). To understand culture as a claim is to understand it as a response to these problems; to recognize it as a tool for coping with the various problems that life presents. While this point will be elaborated in the coming chapters, for the moment I want to flag that understanding human life as something inherently and unremittingly problematic allows us to grasp Jackson's (2005) central point: that culture names the manner in which subjects *manage* (or endeavour to manage) the subject's unrelenting beholdeness to the precarity of living. In this framing, culture offers a means – a resource – to claim one's life as one's *own* life, that is, to claim some sovereignty over one's life even as one is persistently faced with the problem of being a living being – a being that loves (but has no say over its beloved's coming and going), that dies (but cannot choose the time and manner of its death), that opens to others (but cannot predict the consequences of its sharing), that plans for the future (but cannot know what the future truly holds). To understand culture in these terms is to recognize it as a response. It is a system for coping with – though by no means resolving – the condition of being endemically exposed to a life whose parameters, shape, and texture are infinitely beyond our power to manage, manipulate, or control.

There are a number of implications for approaching culture in such terms. First, it alters the direction of cultural theory. For decades, anthropologists, sociologists, geographers, and others have thought of culture as a system of constraint. Whether using functionalist, Marxist, Marxist-humanist, structuralist, interpretivist, or post-modernist frameworks, culture has been thought of as a set of learned systems (norms, ideologies, structures, practices, performances, etc.) that operate as an external force on subjects who could be otherwise. There is obviously a lot packed into this sentence, as the history of cultural theory provides a multitude of mechanisms and outcomes to explain this operation.

But the overall point is that cultural theory – as a project – has framed culture as an external constraint imposed by society on its members. In framing the question of culture as a question of claiming, however, culture is understood as a refuge and a resource; something subjects cultivate and nurture rather than submit to or seek to escape. Given this approach, it is no surprise that culture wars arise when certain identity formations are perceived as under threat. While historically cultural theorists have viewed cultural struggle as a struggle over social hierarchies, the framework I am proposing understands it as a struggle over being – an existential struggle rather than a strictly sociological one.[3]

This leads to the second implication, which is that this theory is not situated in a metaphysics of relationality such as we find in the process ontologies of Deleuze (1987, 1994), Latour (1993), and others (e.g., Haraway 2003). While it does not reject relationality per se, connection is not understood as the defining ontological scene of existence. Vulnerability, precarity, the unresolvable problems of living: these notions make up the primary dimension from which this project proceeds; a dimension that no living being can touch and is simply not available to transformation or becoming.[4] While living beings do have a kind of relationship with precarity, that relation is what Harrison (2007a) terms a relation of non-relation, a relation predicated upon being denied relation. No living being can transform their body's hunger or vulnerability to illness. While bodies can engender relations that feed it or make it well, such actions do not make *the problem* of hunger and illness go away. The problem remains always and unremittingly the same. Thus, even as culture emerges as a positive response to the problem of precarity, it is not a response that has any impact on the problem itself. It is for this reason that I use the term claiming. In understanding culture as something that emerges to manage existential vulnerability, we need to understand claiming in its dual sense. On the one hand, the word refers to something actively staked out or asserted in a positive (if not aggressive) manner, for example, "I claim this territory for France" or "I filed a claim against the company." But the term also carries within it a sense of risk, a contention that is on the cusp of being unwound, for example, "He claims to come from royalty" or "She claims to love me." Claiming in this second sense is an allegation, a desire proffered or mooted but not done so convincingly or without inviting challenge. The point here is that claims to identity and culture must be understood as assertions that are also wagers. While they may be stated positively ("I am a Jew and that means x"), as social scientists we should be wary of affirming those claims as indicative of an abiding interiority. Even as subjects find themselves in an historically bequeathed world with norms, rituals, mythologies, and beliefs already there, this inheritance is not properly *ours*. We, as subjects, have no say over the culture we have been bequeathed. And yet subjects often claim that culture *as if* it were theirs. This is the central question motiving this project: why do subjects not simply live the life they have been bequeathed but claim it – why do they claim it as their own?

The final point to make about this conception of culture is that it is fundamentally material and thus geographical. Rather than approaching the everyday geographical world (the world that is built all around us) as an *expression* of an interior dimension – as something preceded by an acculturated subject (D. Miller, 1994; Duncan & Duncan, 1988) – geography is thought of as the means by which culture (as a claim) is engendered. In other words, geography is given special status here in the constitution of culture. The argument is that there would be no culture if claims did not manifest in material forms. As already suggested, claims have no ontological status. They are evanescent gestures; formations posited in the face of a world that always undermines such efforts. Yet, it is precisely their lack of solidity that summons them in earthly forms. Using the rocks and mortar of our world, claims take shape materially to create something solid and timeless; forming anchors for a life that has no anchor. Exploring culture as a claim, therefore, must begin from its materiality; an approach well-suited to a discipline devoted to examining how the world is written. Indeed, embedded within this argument is a somewhat subversive ambition; a not so oblique suggestion that for cultural theory to survive, there needs to be a certain passing of the torch. It is anthropology that historically approached the question of culture as a question of difference, and it is anthropologists that primarily call for the concept to be forsaken. Framing the question of culture as a claim, however, moves the frame of analysis away from the differences within groups and towards the material gestures they stake. It is the monuments, totems, and skyscrapers of our world that matter; how subjects build their world and how such buildings work to affect culture, not as a modality of presence but as what I will term *a dream of presence*. This is a task well-suited to geography.

This brings us to the title of the book: dreams of presence. If we approach culture not as something a subject has but instead as something outside a subject's ownership or possession, and thus, outside their capacity to narrate or perform in a manner that is reflective of an abiding "self" or any other construct that can be used to describe a self-standing interior presence, then culture comes to take the shape of a fantasy or dream. The phrase "dream of presence" is cribbed from Derrida's essay "Structure, Sign, and Play in the Discourse of the Human

Sciences" (1978); the essay is a commentary on Levi-Strauss's (1969) work on Bororo myth. Derrida's interest is in Levi-Strauss's ambivalence between wanting to elucidate an underlying system for myth – a system that anchors its constant proliferation – and simultaneously recognizing that such a system is itself mythic. The question, for Derrida (1978), concerns the struggle: What is it that requires, needs, or desires structure and foundation? His answer is that the desire is intrinsic to anthropology itself and by extension that which anthropology makes its object: the idea of "Man [*sic*] ... who, throughout the history of metaphysics ... in other words, throughout his entire history – has dreamed of full presence, the reassuring foundation, the origin and end of play" (p. 292). To think culture as a dream of presence, is to recognize that culture, like "Man," does not name or characterize an actual identity. What it names is a dream; not an interior presence, part and parcel of a subject's being, but a desire for presence. Thus, what appears when subjects claim their cultural world is not that world itself: not its presence or its ipseity. What appears is a dream of presence; a fabrication built to claim a world the subject cannot possess. The argument developed in this book is that such dreams must be recognized not as idols to be smashed but as markers that orient our world. They are fabrications because they are things *built* (something manufactured or created) but they are also fabrications in the sense that they are *imaginative* (something consciously and creatively forged). The aim is to acknowledge such dreams (and the relations they establish) as real while simultaneously revealing them to be precarious wonderments, markers of an identity that subjects desire but never actually have.

Arc of the Argument

The previous section was designed to give a basic overview of the argument, illustrating its central claims and its implications. This section dives into the mechanics by giving a précis of each chapter and the overall architecture of the thesis. The book is divided into three main parts, each part representing a specific step in the argument. The first part, "Culture" (chapters 2–4), focuses on justifying the book's opening proposition, that is, that the world subjects are in, and the world subjects claim, are non-correspondent. In doing so, it makes the argument that theorizing culture needs to focus not on the way subjects are different but on the claims subjects make on the differences they embody. The second part, "Claiming" (chapters 5–6), focuses on theorizing the claim itself. Drawing upon Heidegger and Levinas, it uses the dwelling concept to illustrate how subjects struggle to determine some sovereignty – some ownership – over their being. The final section, "Geography" (chapters 7–8), uses this conception of dwelling to rethink what culture *is* and geography's role in making culture manifest. In short, I argue that culture is the means by which subjects claim their being and geography is the means by which those claims become manifest. Below I show how the argument develops in each chapter.

Chapter 2 ("Anthropology and the Naming of Difference") traces the demise of the culture concept in anthropology. The emphasis of the chapter is on the discipline's history in approaching the question of culture as a question about difference. The central point is that the culture concept fails not simply because it cannot help but circle back to various essentialisms (a point well-rehearsed by anthropologists themselves) but also because of certain presumptions embedded in the practice of approaching otherness – the idea that social groups are not simply different but that those differences *belong* to them in some primordial way. Immanent to the ethnographic situation is the presumption that when the native speaks, she speaks for *her* world and her difference. The question I pose is, How do we approach such claims? Through a discussion of anthropology both past and present, the chapter illustrates the importance of dividing the question of culture into two questions: a question about difference and a question about claiming.

Chapter 3 ("Geography's Cultural Legacy") turns its attention to the discipline of geography and its own history of theorizing culture. Specifically, I argue that there are two traditions of approaching the question of culture in geography: an Anthropogeographical School and a Representaional School. While the former focuses on human-environmental interactions (and the habits it generates), the latter explores how subjects come to invest in those habits as things that are both meaningful and theirs. The bulk of the chapter is dedicated to fleshing out these two traditions, but the chapter concludes by arguing that both the question of habit and the question of meaning are necessary for understanding culture. While it is important to understand how certain forms of life take shape historically through relational interaction, there is a second question here about why subjects come to invest in or claim those ways of life as their own. Again, the question here is a question of claiming: Why don't subjects simply live the lives they have been bequeathed? Why do subjects claim them as theirs?

Chapter 4 ("Consciousness as Claiming") concludes the section by examining what I term the *practice-habit* literature in geography and anthropology and the way it understands self-consciousness. The central argument is that while conceptions of habit and material practice provide compelling accounts for explaining difference, their understanding of why subjects attach meaning to those habits is more limited. Drawing upon the psycho-analytic work of Jean Laplanche, the chapter argues that self-consciousness is not simply about being aware, it is also about being self-possessed. In this framing, consciousness emerges as a desire to claim oneself as a sovereign subject in the face of various existential problems that exceed our practical capacities. It concludes by illustrating how the question of claiming provides an original basis for rethinking the question of culture.

As previously suggested, section 2 moves on to theorizing the claim itself. It does this through an engagement with the concept of dwelling as it arises in the work of Heidegger (chapter 5) and Levinas (chapter 6). The dwelling concept is central because it presents human subjectivity as caught in an existential struggle to create a place for itself – a site where it can be a self-standing, self-possessed being vis-à-vis the various cultural, historical, existential forces that define its world. In this sense, it provides the basis for thinking subjectivity as an endeavour – a struggle – to claim oneself as sovereign. That said, Heidegger and Levinas have very different frameworks for understanding this struggle and its consequences.

For Heidegger, dwelling is conceptualized as a struggle between the subject (Dasein) and the world that defines its being. The central contention of chapter 5 ("Heidegger's Dwelling") is that dwelling marks an ongoing attempt in Heidegger's work to mitigate or temper his central philosophical claim, that subjects are wholly in-the-world. For Heidegger, subjectivity must be thought as a reflection or modulation of the world into which subjects are thrown. Yet the problem with such a framework is it provides very little room for human creativity and agency. Dwelling is how this dimension of humanity gets thought and it is a concept that evolves significantly from his early to later work. While Heidegger's *Being and Time* (1962) presents dwelling as one of many metaphors for everyday existence, his later work uses the term to describe how Dasein has the capacity (indeed the obligation) to build its world creatively using the resources it finds. It is in this manner, Heidegger argues, that Dasein establishes some sovereignty over its own existence, that is, creates a space for itself – a place that is its *own*. In addition, while Heidegger understands such creations to be perpetually and unremittingly undermined by the world itself, the dwelling nonetheless names a desire, an ambition, within Dasein to establish one's self as a sovereign being.

In chapter 6 ("Dwelling as Both Open and Closed"), the discussion turns to Levinas and explores his own unique interpretation of dwelling. At a basic level, Levinas understands the dwelling in similar terms to

Heidegger: a place or site where the subject seeks to claim some sovereignty or ownership over its being. But while Heidegger understood the subject's sovereignty as threatened by the unfolding nature of the world, Levinas understands it as threatened by the subject's responsibility to the Other. In the former, Dasein's struggle to dwell is cyclical. Dasein builds a place for itself and the world undermines such buildings. In the latter, however, the dwelling is never properly built – it can never be properly built. This is because the Other can never be shut out. As Levinas suggests, the dwelling is a site that is always both open and closed. While subjects need closure in order to think themselves as sovereign beings, they can never achieve this closure because their existence relies on being open to the Other. While the chapter spends significant time explaining this contradictory relation, suffice to say for the moment that it creates an impossible situation. The dwelling, in Levinas, is not something that can ever by achieved – not even in a momentary or cyclical way. On the contrary, it can only be pursued. This is to say that sovereignty and self-possession, in Levinas, never properly takes place. It is only ever a claim.

This brings us to the final section. Section 3 (geography) develops this conception of claiming into a broader theory of culture. Chapter 7 ("Culture as Claiming") focuses on reinterpreting Bruno Latour's notion of cultural ontologies. For Latour, culture can be conceived as ontological in the sense that it constitutes a fragile relational system of knowing which provides practical solutions (what he calls "paths") for navigating a hyper-active and unstable world. Yet while Latour conceptualizes this ontology as a system of forces and relations, this chapter reframes it as a response to existential alterity. This places Latour's notion of cultural ontologies in a very different light. Rather than being an always contingent architecture of forces, cultural ontologies are understood as desperate desires; not modes of existence but modes of response. While they are still learned paths, they operate not in a world of relations but in a world that denies relation; a world whose parameters are not always available for composition, enrolment, or force. The second half of the chapter develops this framework through an engagement with Michael Jackson's conception of existential anthropology; an approach that conceives culture as a response to universal existential conditions rather than as a system of distinctive sociological structures.

This leads to the last substantive chapter: chapter 8 ("Fundamental Geography"). The aim here is to develop the idea that culture is a representation. While the concept of representation has been thoroughly discredited in recent years, this critique was predicated upon the presumption that representation is something done by subjects in order to achieve some political goal. The argument here is that representation is how subjects come to exist, that is, it is the means by which subjects mark themselves as self-standing beings in the world. It is only through representation that subjects can conceptualize themselves as sovereign beings. In addition, it argues that geography is fundamental to this practice. While subjects can potentially mark their being through various technologies (writing, language, signs, etc.), it is the material world – the bricks, mortar, fabric, and clay of landscapes and environments – that provides the most potent source for making our life appear as something stable, enduring, and existent. Geography is thus not simply one means among many. It is the most significant way that human subject's mark their lives. It is by writing ourselves into the earth that culture comes to appear in the world.

The final chapter, chapter 9 ("Dreams of Presence"), draws the argument to a close by illustrating how our everyday material geography does not reflect or express internalized identities. Rather, it is the means by which the dream – the fantasy – that subjects *have* identities appears. The chapter argues that what we see in our everyday material world is not culture's presence but a dream of presence. It is by writing ourselves into the material world that subjects come to identify not simply who they are but – more fundamentally – that they exist. Throughout the text, I have argued that subjects have no stable interiority, no anchor to which their being can be fixed. What subjects do have, however, is geography. It is by building the world into recognizable material structures that subjects mark themselves to themselves and to others. Understanding culture as a claim means recognizing identity as a fragile and evanescent thing – never assured, never present. But it is through geography that subjects dream themselves as present. It is through representation that subjects claim their existence – their presence – as their own.

Conclusions

I began this introduction by suggesting that the culture concept has gone out of fashion, its territory ceded to other conversations and frameworks, many of which are themselves disingenuously caught in the same problematics anthropologists and others debated many years ago.[5] Yet I have also argued that a reassessment of the concept is well overdue: while the war over culture may have been resolved in the academy, it has by no means been resolved in the world. The loss of culture as an organizing conceptual tool has not been without consequence. As the contemporary political landscape appears increasingly defined by battles over identity, nativism, heritage, and culture, we seem to have precious few concepts for making sense of these dynamics. And yet how to avoid going around the same circle, skirting the eventual fall into various essentialist anchor points?

My proposal is to change the question of culture from a question about difference to a question about claiming. To be sure, I am not the first to suggest that the answer to a theoretical dead-end lies in a reconfiguration of the question. But the reconfiguration I am suggesting here is not simply a switch from one object of analysis to another. It is also a switch towards a certain mode of theory. For the last twenty years, the purpose of theory has been to expand our purview away from human-centric thinking. The process ontologies of Deleuze (1994), Latour (1993), Haraway (2016), and others (e.g., J. Bennett, 2010) have been instrumental not only for introducing a host of new concepts, interests, and objects but also for repositioning the human as an effect of a more generalized metaphysics of becoming; a framing that situates the human as an emergent compositional event predicated upon affects and encounters that are not unique to human life. While such ontologies have been very good for understanding the open-ended, becoming nature of (human and non-human) existence, they are poor at recognizing the singularity of any particular existence. In their enamourment with *life*, they have forsaken the capacity to consider *a life*; a life we mourn because its ending rips a hole in the world; a rupture that is not simply retied through new relations but remains irreparable (Harrison, 2008, 2015; Philo, 2017). While this book does not constitute a critique of process ontologies (cf. Bissell et al., 2021; Harrison, 2021), its emphasis on claiming signals a very different theoretical orientation. Rather than focusing on how things exist, it explores how certain modes of existence come to be thought, understood, and claimed as *mine* – something that is uniquely and singularly understood as my own. This conceptual universe is not defined by the question "What makes me a subject?" but rather "What makes me *not an object*?" – something more than the habits and routines that my body expresses, something more than my difference.

It is only when we understand the question of culture in these terms that we can recognize that the struggles over identity and meaning we see in the world today are not simply battles, they are wars. Culture wars are not struggles over status, hierarchy, and power. They are about *existence*. This is not to suggest that power relations are not implicated in these struggles. The point is that their instigation is existential before it is sociological. The fact that Christian nationalism may be an incoherent set of beliefs, eschatologies, practices, and policy positions, does not detract from the fact that it provides a means for naming a distinctive mode of life. And if that terminology is threatened or undermined by other identity narratives, which are themselves constituting various subjective capacities, then it is hard to see what other options exist besides war. As Žižek (2002) argues, self-hood is not a matter of knowledge – it is a matter of faith. While I may know that my self-identity is mobile, fluid, relational, and becoming, I nonetheless have faith that it is *mine*. And while politics is negotiable, faith is not.

Thus, while this book is a theoretical book, it is not theory for theory's sake. And even though its arguments may seem far from urgent issues of race, identity, power, and injustice, this is only because it moves towards them from an opposite point on the circle. It is only when we understand culture and identity as a claim that we can begin to think about the politics of these claims and how to take them seriously. To be sure, this text does not focus on these political questions enough and that absence will no doubt be an issue of consternation for some. By way of pre-emptive defence, I will only say that this book is already doing a lot and while I have a number of thoughts about how the argument here can be applied to the political concerns of our day, those thoughts will have to wait for another book. My more immediate ambition here is to suggest that these issues cannot be addressed sufficiently until we have a working theory of culture. As Oswin (2020) and others (e.g., Bonds, 2020; Noxolo, 2017) have argued, despite the academy's recent efforts to decolonize the curriculum and address social and political inequities related to race, class, and gender, there is still very little actual research in geography on these topics. One of the reasons for this, I would argue, is that we still do not know how to *think* about these things, or at least think about them in a way that does not devolve into various essentialist or structurationist rabbit holes. My hope is that this project can potentially point a way forward on these problems, even if I am not the one to lead the march (or at least not yet). Indeed, while the immediate plan and purpose of this introductory chapter is to introduce the argument's basic terms and positions – all of which still need to be clarified, justified, and defended – its broader and more modest aim has been to make the point that the question of culture still matters. And to illuminate what has been lost in forsaking it. Whether this justifies a book-length treatment on the concept, or provides any convincing remedies, waits to be seen. But even if the arguments fail to satisfy, my hope is that they will at least help us see the stakes involved. Culture wars are a consequence of culture's loss. And unless we take seriously the existential problems at the heart of culture, we will continue to be deaf to the forms of faith (including our own) that instigate those wars and indeed our own investment in their perpetuation.

SECTION ONE

Culture

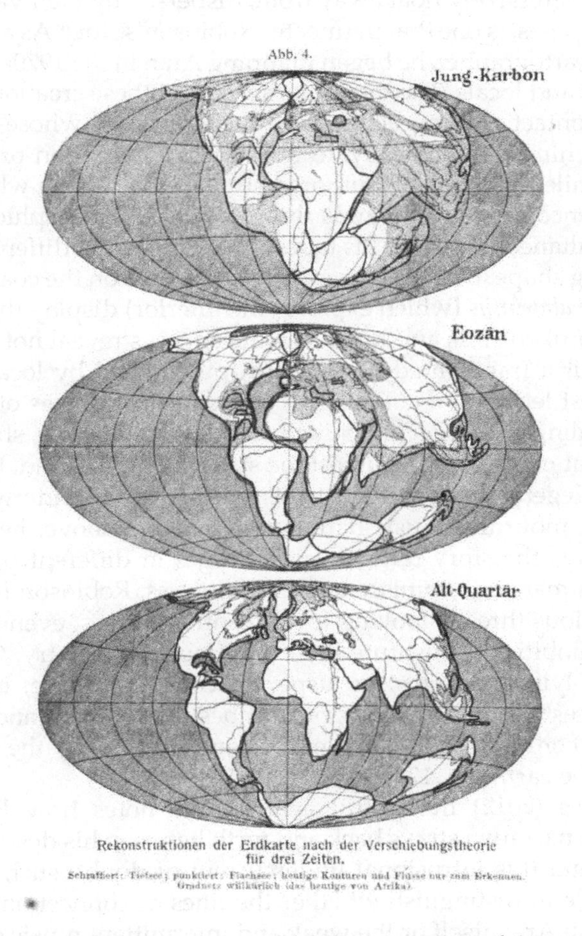

Figure 1. Original world maps created by the progenitor of the continental drift theory, Alfred Wegener.

Dreams of Sameness

In Tim Robinson's *Stones of Aran: Pilgrimage* (1986), the first of his two detailed studies of the Irish islands, he opens with a short discussion of geological time and the politics of continental drift: "The geographies over which we are so suicidally passionate are ... fleeting expressions of the earth's surface," he states. "Two hundred million years ago the Atlantic did not exist and all the land-masses of today were clasped together in one continuity, in pre-Adamite innocence of the fact that one day scientists inhabiting its scattered fragments would give it the lovely name of Pangea" (p. 7). This image of a shattered world, a lost unity that humans progressively float away from, dispersed by their various continental voyages, is one that permeates Robinson's story. As a mathematician and cartographer, he began mapping Aran in the 1970s, primarily for tourists and locals. The moderate success of these creations brought him into contact with specialists and local villagers whose input and stories led him, paradoxically, to question the ambition of his initial project. While the maps he authored strove for a certain wholeness, a means to encompass the islands through their cartographic reach, he could simultaneously sense his tapestry fraying and a different kind of story taking shape. Both *Pilgrimage* (which focuses on the coast) and the follow-up *Labyrinths* (which explores the interior) display this restlessness. The broken cliffs and headlands he surveys reveal not a contiguous line but a fragmented topography, punctuated by local lore and legends, lost legacies of colonialism and daring practices of climbing, fishing, scaling and collecting suited to the promontories, shelves, and crevices that criss-cross the limestone surface of the island. Commenting on the legend of a piper who gets trapped in an underwater cave, and whose mournful music haunts the villagers above, he notes the various ways the story refracts and changes in different parts of the island. In a manner reminiscent of Levi-Strauss, Robinson follows the myth's various threads looking for its central themes, even as he realizes such ambitions constitute their own mode of artistry: "The artist finds deep-lying passages, unsuspected correspondence, unrevealed concordances" Robinson states, but "at best it is a weak and intermittent music confused by its own echoes and muffled by the chattering waters of the earth" (p. 129).

In Wylie's (2012) thoughtful analysis, he notes how Robinson's ambivalent narrative strays back and forth between his desire for completeness and this intrusional undercurrent of doubt, such that he is often unable to distinguish whether the lines of connection he draws emerge from Aran itself or the weak and intermittent music of his own making. The irony, as Wylie suggests, is that the deeper Robinson digs into Aran, the less the island coheres: "The more [Robinson] says ... the more Aran ... withdraw[s] from view" (p. 378). Robinson's map is one that resists its own ambition, perpetually breaking its own imagery, like the currents and wind that erode its surface.

When I first read Robinson's books, I was oddly reminded of my obsessive consumption of ethnographic studies at university. Even though I was not an anthropology major (I had little interest in physical anthropology or linguistics), I availed myself of the department's course offerings and read any ethnographic work I could find on the tribes of Central Asia and the Levant. While I cannot remember much about the titles or the specific content of these books, I do remember my ambivalence when checking them out of the library. On the one hand, I thought they would be worthy reads given my interest in the history, culture, and languages of the region. But I also found them boring. With their long lists of rug-making styles or shelter-building techniques, the books seemed to aim for a certain comprehensiveness, as if the whole culture could be catalogued within its pages. Later, as a postgraduate student in geography, I came across a very similar mode of writing in the regional studies promoted by Hartshorne (1939), James (1942), and Hart (1982) where (once again) completeness seemed to be an aim in and of itself. "Geography seeks to acquire a complete knowledge of the areal differentiation of the world," Hartshorne (1939) boldly states, a "total differentiation of areas" (p. 463).[1] Through Robinson's eyes, I began to see such exercises in list making in a different light; not necessarily as knowledge but as artistry, their own pangeac dream.

The aim of this section is to illustrate how the tradition of approaching the question of culture as a question about difference (how others are different from us), embedded a series of contortions on the problem that effectively made the question impossible to answer. While each chapter approaches this issue through a different literature, collectively their aim is to illustrate how framing the question in such terms makes the problem irresolvable. Difference, I argue, defies knowability. Regardless how exhaustive one's list, how detailed one's encyclopaedic ambition, difference defies what we can say about it. This is precisely the problem with approaching difference as an intellectual problem. While difference is everywhere, obvious and incontestably present, it resists explanation. Any and every ambition to define, understand, and explain difference always and inevitably becomes an exercise in essentialism (Dulley, 2019). This is not to deny that there are profound differences that distinguish communities, places, and groups. Indeed, I would argue that such differences are real, abiding, and even essential.

And yet, they are also unknowable. They exist but cannot be named. They are obvious but defy explanation. The more we write about difference – the more we list its attributes and map its boundaries – the more it withdraws from view.[2]

The three chapters in this section illustrate this point through three different traditions: cultural anthropology, cultural geography, and what I am terming the *practice-habit* literature. In each chapter I illustrate how the question of culture is framed as a question of difference and how – in doing so – the arguments about culture struggle to resolve themselves. I do this even when the terminology of culture is not used, for example in the ontological turn literature in anthropology or in the atmosphere literature in geography. The aim is to illustrate how the endeavour to name difference is a pangeaic dream, a dream of sameness. These disciplinary summaries no doubt embed their own essentialisms and will undoubtedly be contested by their practitioners. But the argument about the problem of difference itself – the problem that difference imposes for theory – will be more difficult to dismiss. In establishing the problematic nature of difference (problematic in the sense that it is irresolvable), my ambition is to change the question of culture from being a question about difference to being a question of claiming. The chapters justify this pivot and, in doing so, set the stage for the next two steps in the argument.

Chapter Two

Anthropology and the Naming of Difference

Anthropologists have created considerable difficulties for themselves in their numerical vision of a world full of societies and cultures.

– Marilyn Strathern, *Partial Connections* (1991)

Anthropology's Otherness

As suggested in the previous chapter, it would seem irresponsible, if not disrespectful, to talk about the culture concept without engaging with those knowledges, literatures, canons, and conversations associated with, or self-identifying as, the discipline of anthropology. No community of scholars has engaged with the question of culture for so long and so diligently. No intellectual tradition has taken the question of difference (and human difference, in particular) so seriously and so openly. To read anthropology is not simply to step into a different discipline with different traditions, emphases, and tweaks. It is to step into a different world. Journal articles are never what they seem. With titles that promise to talk about gender relations, epistemology and class conflict, they quickly become steeped in detailed conversation about Nuer kinship patterns, Chewong economies, Maori gift exchange, or Melanesian cosmologies with the point in the title delayed until these discussions are fully developed. While geographers tend to use empirics to support their theory, anthropologists often think theory through their empirics; the event of otherness itself being the instigation for thought.

To be sure, this rudimentary gloss of an entire disciplinary tradition is both totalizing and essentialist. It is totalizing because, as Viveiros de Castro (2004, 2014) regularly reminds us, there are many anthropologies, that is, many ways of seeing otherness and many ways of naming it. It is essentialist because anthropology cannot be reduced to a particular way of writing, thinking or encountering otherness. On the contrary, the difference of anthropology (and the differences within anthropology) are unfolding problems; problems that I (as a geographer) am struggling to describe and problems that anthropology, as a set of institutional traditions, interrogative habits, inscriptional formulae, and methodological conventions, has dedicated itself to proliferating. Indeed, one could argue that the problem of talking about anthropology's difference is precisely the problem of anthropology itself. The problem of naming what is different about others. This brings us to the aim of the chapter. On the one hand, the aim of this chapter is to illustrate the problem of identifying, studying, describing, and conceptualizing difference. In particular, it is to illustrate the impossibility of the problem and how its impossibility has led to the culture concept's demise. To be clear, this is a problem that anthropologists themselves have narrated many times (Clifford & Marcus, 1986; Kuper, 1999). And yet, it is not only necessary terrain for a book such as this – a book that is dedicated to reconstituting and rehabilitating the culture concept – but also, I would argue, a problem that has not yet been resolved within the anthropological tradition. Indeed, while anthropologists have, for the most part, ejected culture from their conceptual vocabulary, the problems associated with naming difference continue. This leads to the second aim of this chapter, which is to outline an alternative proposition for approaching the question of culture; an approach that is not about difference itself but about claims to difference. Here, I argue that it is only by separating the question of difference from the question of claiming that the culture concept can be moved to a different theoretical ground; a ground that allows us to approach difference not as something knowable but as a claim whose parameters can be researched without the danger of essentializing or ontologizing its reality.

The discussion is divided into three parts. Part one focuses on exploring some of the theoretical dynamics that led to the culture concept's demise in anthropology. Drawing upon the work of Argyrou (2002), the section illustrates how conceptualizing culture in terms of difference (as a term that encapsulates the differences between groups) ultimately led to the concept's dissolution. As Argyrou suggests, to think culture primarily in terms of difference is to think of it as something that other groups have. Thus, even as anthropology understands itself as a discipline dedicated to understanding humanity in all its variation, theoretically the culture concept works by naming and explaining the difference that it sees – difference as a phenomenon witnessed (empirically) in others. While the implicit Eurocentrism of this view has been a topic of much debate (see R. Wagner, 1981; Viveiros de Castro, 2014),

I am more interested in the relation of propriety such an approach establishes. For in asking the native[1] to speak for her difference, she is summoned to speak for a world, a mode of life or a form of existence, as if such thing were *hers*. In this sense, it is not simply that we understand the native as different but that we understand that difference as something that belongs to her; something that she possesses, and in some fundamental way can speak for.

The second part of the chapter endeavours to illuminate how this dynamic of possession continues to inhere in contemporary theoretical trends in the discipline. Drawing upon work often associated with the ontological turn (Holbraad & Pedersen, 2017; Strathern, 1991; Viveiros de Castro, 2014), I make the point that ontological approaches do not get around the problem of self-possession. For thinkers such as Viveiros de Castro, Indigenous concepts are ideas thought to be immanent to Indigenous worlds. Thus, while the interests and goals of the ontological movement are distinct from traditional cultural anthropology (particularly work interested in culture and cultural difference), inherent to the idea of native ontologies is the presumption that certain modalities of knowledge *belong* to the native. As in the previous section, my aim here is to both illuminate how the phenomenon of self-possession continues to appear in anthropology and question the correspondence between the claims that appear and the world that is claimed. This leads to the final part of the chapter, which explores how the question of claiming constitutes a distinctive phenomenon upon which a novel question of culture can be built. Here, I draw upon Tim Ingold's (2000, 2007, 2008) work to illustrate how the question of culture is actually two questions: there is a question about difference (i.e., about why certain patterns of difference appear, whether they be in bodies, affects, networks, relations, etc.), and there is a question about why subjects claim those differences (why do subjects claim difference as something that is theirs, i.e., that belongs to them)? The chapter concludes by arguing that this second question of culture – the question of claiming – is fundamentally different from the first, and thus, needs to be addressed on its own conceptual terms. In this sense, it presents a distinct ground upon which to build a theory of culture.

Culture as Cultural Difference

It has been over thirty years since anthropologists began critiquing the culture concept as a tool for making others different. As Abu-Lughod (1991) puts it, "As a professional discourse that elaborates on the meaning of culture in order to account for, explain and understand cultural difference, anthropology also helps construct, produce and maintain it" (p. 143; see also Kuper, 1999). Culture, in other words, is a concept used by anthropologists to create their own object. In the very endeavour to identify and name other peoples' difference as culture, the concept essentially "constructs" (to use the parlance of the time) the difference it seeks to understand. Abu-Lughod (1991) was not the first to make these arguments and, in many ways, her critique reflects a broader concern throughout the discipline with the writing and representation of others. But despite the powerful contortions these critiques levelled against the discipline throughout the 1990s (critiques, it could be argued, that first led to the concept's inevitable demise), the problems they exposed ran deeper than how to write about and represent others. As Argyrou (2002) suggests, in his extensive and wide-ranging critique of anthropological thought, postmodern anthropology never really problematized the equation of culture with cultural difference. What it critiques is the idea that anthropologists can represent difference, that they can leave behind their own epistemological categories in order to access (and represent) the difference of others. But what is not put into question is the equation of culture and difference itself. In other words, postmodern critiques do not deny a locatable (and potentially knowable) difference in others. Difference is still very much *there*. It is just hidden, locked away behind an epistemological wall of otherness, "a reality wrapped up in itself, forever inaccessible to us" (Argyrou, 2002, p. 30).[2]

This equation between culture and cultural difference in anthropology is the focal point of this critique and, as such, it piggybacks on Argyrou's (2002) central argument concerning the history of anthropological thought. For Argyrou, the underlying problem haunting anthropological theory is that difference is understood (always and unremittingly) as a problem: a problem that the anthropologist needs to solve, explain, or address in some manner. This is because, Argyrou argues, anthropology begins from a presumption of human sameness. It is only because humans are understood to be essentially the same that difference appears as an *interruption* or perturbation of this primordial norm. Difference appears as a phenomenon because it stands out from sameness. And it is precisely this standing out that leads it to be thought of as a conceptual concern, that is, as a phenomenon that needs to be addressed. Thus, culture as a concept works for anthropology because it gives the discipline a question to answer. It establishes culture as a problem (the problem of difference) and directs anthropological thought towards its resolution.[3] This is why despite anthropology's long history of ideas, frameworks, approaches, and epistemologies, Argyrou argues that all such efforts work towards the same goal: explaining a community's deviation from and/or modification of what he terms *the Same*. It also explains why the culture concept cannot escape its configuration as a term to explain difference. Culture, from the beginning, is invented to explain a phenomenon already deemed to be a problem – the problem of difference.[4]

There are two implications that follow from Argyrou's (2002) argument that I want to trace out. First, in equating culture with cultural difference, culture is not simply reinforced as anthropology's object, it is also rendered as an object that is *knowable*. Thus, culture not only provides the means to name difference, it also promises to reveal the mechanism behind that difference, the underlying forces working to modify sameness. It is for this reason that Argyrou argues that all theories of culture must terminate in the Same. In explaining how these knowable mechanisms work to engender a particular cultural group's difference, they are explaining that which disrupts sameness (which is *apriori*). This leads to the second implication, which is that conceptualizing difference as a knowable problem establishes a proprietary relation between difference itself and the other people that express it. In other words, it is the conceptualization of difference as a knowable problem that leads anthropology to position the native – the other who is steeped in that difference – as the key to unlocking that knowledge. Hence, the emphasis on ethnographic work. It is only by asking the native to account for herself – to explain and describe her difference – that the anthropologist can understand. "The trick," as Geertz (1983) once put it, is to "figure out what the devil they [the native] think they are up to" (p. 58). It is only by understanding how the native herself understands her actions, behaviours, and inclinations in regard to her difference that that difference can be explained. The trick is knowing what she already knows. Because, as Geertz make clear, "no one knows ... better than they do themselves" (p. 58).

It is this second implication that is the emphasis of my own critique of the culture concept. The argument, to be clear, is not that the anthropologist presumes that the native knows herself transparently or that she can speak about her world with eloquence, felicity, and self-reflection. My point is that the very project of asking the native to account for her difference, in positioning her as best placed to narrate and explain her world and what (the devil) she thinks she is up to, the anthropologist posits an interior dimension from which the native is thought to speak. An interior dimension that is hers and that corresponds to the world she is in. To ask the native about her difference is to presume that she *can* account for her difference – that she knows something about it because it belongs to her in some primordial way. Indeed, while Abu-Lughod (1991),

Ortner (1999), Clifford (1983, 1986), and others are very good at questioning the anthropologist's capacity to know the native, they never properly question the capacity of the native to know herself. This is because – as Argyrou (2002) illustrates so well – a primordial sameness situates the native within her difference. There is a correspondence that readily flows from her to her difference. The native belongs to her difference and her difference belongs to her.

It is this sense of ownership and belonging – this idea that one's culture is one's own and thus something we (as native or anthropologist) can speak for – that I am attempting to bring into question. While I do not deny that the native has a point of view (don't we all), the problem comes when we draw links from this point of view to her difference. Thus, even though the native no doubt speaks and may indeed speak eloquently about her difference, it is the presumption of sameness that allows us to see that difference as belonging to her, as if the native has some ownership, insight, or command over the difference she has inherited. To be clear, my critique here is not about questioning whether other people (native or otherwise) are different. On the contrary, difference is everywhere. My critique concerns what I term the *proprietary gesture*, the presumption that difference is not simply there but is *mine*, that I belong to it and can speak for it. Rather than understanding the native's difference as something she has been given (something she has no say over or even privileged knowledge of), we presume a correspondence. Difference (as an aberration of sameness) is knowable. And no one knows it better than the native herself.

To conclude this section, the point here is not to level an assault on anthropology or the anthropological project – or at least not one that has not already been made by its own practitioners. It is to illustrate that the problem with the question of culture is not about the question itself but about the way that the question has been framed. In presenting the question of culture as a question about difference, the concept of culture is bolted to a desire to understand the differences inherent to subjects (Viveiros de Castro, 2004b). Even as theorists such as Clifford (1988), Abu-Lughod (1991), Kuper (1999), and others (e.g., Gupta & Ferguson, 1997) have rigorously undermined the idea of the cultural subject, their theories nonetheless remain predicated on subjects that *possess* different identities, even if we were not willing to name culture as the abiding form those identities take. Clifford (1988), for example, who positions culture as a serious fiction, nonetheless sees it as a fiction written by a subject about their difference. The process he terms *fashioning of the self*, while no longer moored to a social, territorial, or historical legacy, is nonetheless predicated upon a subject's capacity to enrol a multitude of diverse signs, references, and representations, operating on a global scale, into distinctive identity expressions – expressions that are conceived as their own (see also Battaglia, 1999). The subject, in other words, retains its capacity to possess its identity and testify to its difference. And it is precisely this capacity to possess that I am bringing into question.

Ontology and the Naming of Worlds

The aim of the previous section was twofold. On the one-hand it was to illustrate why the question of culture went into decline. When framed as a question of cultural difference, the concept cannot help but conjure up the shadow of the same as well as the spectre of a self-present subject who speaks for a world that is hers. On the other hand, it was also to suggest that if the question of culture could be framed differently, that is, in terms other than difference, then the culture concept could be potentially restored. In order to start this process, I want to return to the proprietary gesture I drew out in the previous section. To be clear, the critique I made above was not about doubting the dynamic of possession as a phenomenon. That would be ridiculous. We and others speak for our world, our culture, our community, our history, and we speak for them (as the possessive pronoun repeatedly suggests) as things that are ours, that is, as things that are part of us. The target of the critique in the previous section was on the presumption of correspondence, the idea that the world the native is in and the world that she speaks for *are the same*; that the native knows and understands the world which she is asked to testify for because that world is *hers*. As suggested in the introduction, it is this claim to correspondence that interests me, and it is upon this phenomenon that I am seeking to develop an alternative approach to the question of culture.

In order to do this, I want to first illustrate how the claim inheres in more recent modes of anthropological thought, specifically work often associated with the ontological turn. Even as ontological anthropology (and there are a number of different flavours of this) presents a significantly evolved conception of subjectivity and lays out a very different kind of anthropological project, the problem of correspondence continues to haunt in expected and unexpected ways. This is despite the fact that many of the arguments I make in the previous section are echoed in various corners of this work (see Viveiros de Castro, 2014). And yet the aim of this section is to illustrate that in certain expressions of ontological anthropology the proprietary gesture abides. In other words, there remains an inclination to associate certain people with certain worlds; worlds that are not simply there but understood to be (by native and anthropologist alike) their own. The aim here is to illustrate how this problem continues to reappear in various facets of ontological anthropology, even as this movement points to some promising directions for reframing the culture problem.

The body of work often associated with ontological anthropology (for lack of a better term) is hard to summarize, particularly as it has come to constitute an ever-expanding realm of debate. Holbraad and Pedersen (2017) identify two distinct strands. One that draws upon alternative (often Indigenous) ontologies in order to be experimental and reflexive, characterized by the work of Evens (2008), Kohn (2013), R. Wagner (1981), Blaser (2009, 2014), De la Cadena (2010, 2015), and Viveiros de Castro (2004a, 2014), and a second which endeavours to excavate the ontological worlds of others, which is characterized by the work of Sahlins (1985, 1999), Scott (2007), Descola (2013), and Descola and Gaisli (1996). For this review, I am going to focus on the former at the expense of the latter. The rationale being that the latter is more obviously invested in naming particular modalities of existence as "ontological" in its traditional Heideggerian sense. For anthropologists such as Scott (2007) and Descola (2013), the communities they study exist within distinctive ontological worlds that can be named and situated within particular geo-historical locations. As Holbraad and Pederson (2017) suggest, "this way of 'locating' ontologies … reiterates, albeit at a different ontological 'level,' more standard ways of thinking about, say, cultures" (p. 60). In other words, while there are certainly important distinctions to be drawn between the concept of culture (in its traditional epistemological guise) and the concept of ontology (as it is being advanced in this realm of work), the inclination to associate particular modalities of life with distinctive conceptual formulations of that life, certainly creates an opening for the critique Argyrou (2002) develops above, or in Holbraad and Pederson's (2017) words, "if ontologies, like cultures, are objects to be found out there in the world … then the spectre of all the discomfiting questioning to which the notion of culture has been subjected in the past half-century returns" (p. 60).[5] The point of this review is to illustrate how the first strand of ontological thought *also* falls into this problem – though through a far more circuitous route. In order to illustrate this, I would suggest that this trajectory of thought can be broadly summarized through three characteristics: (1) a critique of epistemological approaches to otherness, (2) a relational conception of subjectivity, and (3) an interest in native conceptual resources and ideas. I will discuss each of these in turn and then develop the critique.

In terms of the first characteristic, the works of Argyrou (2002), Viveiros de Castro (2003, 2004a, 2004b), Holbraad (2010, 2007), Povinelli (2002), Ingold (2000), Pedersen (2011), Strathern (1991), and R. Wagner (2001) have all put forth sustained critiques of epistemological conceptions of culture. While there are important distinctions to be made in these works, broadly speaking, there is an attempt, as Henare et al. (2007) suggest, to move the conversation away from a discussion of world views in order to consider the nature of worlds. In the terminology of the former, culture is an epistemological matrix that overlays an empirically verifiable reality. The question of culture, in the language of epistemology, is a matter of the mind, an internalized modality of perception and understanding that colours or frames the way subjects view and/or represent an external reality. The critiques that the above-mentioned authors develop are extensive and elaborate. For Ingold (2000), the position grants a total lack of agency to the non-human; for Viveiros de Castro (2003), it promotes an intersubjective hierarchy between the scientist and the native; for Strathern (1991), it equates all cultural representational cosmologies as equal and substitutable (see also Argyrou, 2002); and for Henare et al. (2007), it reinforces a positivist ontology that wedges a strong division between internal (i.e., representational) and external (i.e., factual reality) worlds. In response, the authors moot the existence of different worlds where the "defining problem consists less in determining which social relations constitute its object," instead "asking what its objects constitutes as a social relation" (Henare et al., 2007, p. 4; see also Viveiros de Castro, 1998, 2004a). In other words, the concept of a world stands for a field of possibilities where the very notion of what a subject, what an object, and what the relation between them *is* can be thought (Viveiros de Castro, 2003, 2004b). Worlds, in this view, are not simply economies of meaning and representation but economies of life itself, that is, economies where the very projects that define what living is, what existence is, and what modalities for existence (as a mother, a warrior, a shaman, etc.) are possible, defy and challenge the anthropologist's pre-conceived categories of the real. In this sense, there is no strong distinction between the world that is out there and the world of the subject, not because there is no world out there (indeed defending radical idealism simply reiterates the epistemological priority of previous theories in different terms), but because the world being described is of far greater depth and dimension than what positivist conceptions of reality and truth allow. Responding to the questions it is asked, different worlds give data to the scientist, as much as they give guidance to the shaman.

The second feature of ontological approaches is their approach to subjectivity. While epistemological approaches understand subjectivity as mobile, performed, contextual, and so on, there remains within this work a subject defined by an abiding interiority; a site within the subject that (to varying degrees) provides a basis from which they can account for their acculturated self. Drawing from the process ontologies of Deleuze (1994), Latour (1993, 1997), Haraway (2003), and others (e.g., Serres & Latour, 1995), there is broad consensus that subjectivity is an effect of its relations. Rather than being a self-standing, self-contained being, defined by its interior point of view, the native's view is an effect of the relations in which it is situated and unfolds in tandem with those connections. Thus, while understanding and comparing the native's differences remains a cornerstone of the ontological project (Holbraad & Pedersen, 2017; Viveiros de Castro, 1998, 2004b), how that difference is understood is significantly transformed. Difference here is not a difference of culture – that is, a difference of otherness wrapped up in an acculturated self. There is no abiding interiority – no abiding sameness within the native – that defines her or characterizes her world. On the contrary, what the native speaks for is her unfolding relational situation. The relations *precede* her and she is an expression of their emergence. It is for this reason that Strathern (1991) suggests that the anthropologist must "turn the mind with no individual locus into a much stronger sense of exteriority" (p. 26). By this Strathern means we need to look less at the native in front of us and more at the presence of others – other beings, forces, events, and problems – constituting the native and her world. Strathern writes, "One needs to restore a perception of other presences, those who jostle, pressing in, as concrete and particular others who will neither go away nor merge with oneself" (pp. 26–7). The point, for Strathern, is that anthropologists need to stop looking for the native's soul and start looking at their field of relations. Difference, and the fractal processes by which difference unfolds, is the generative heart that allows the native and native's world to be seen. It is the presence of exterior others (those that jostle and press in) that constitute her. She does not precede these relations but is an effect of their unfolding. The native emerges from that which is exterior.[6]

The third characteristic, and perhaps the most innovative aspect of work associated with the ontological turn, is the novel conceptual project it sets. Taking a page from Latour's (1987, 1988) empirical philosophy – or perhaps from Deleuze's (1990) transcendental empiricism – ontological anthropologists advocate a new modality of doing anthropology (Laidlaw, 2012). As Carrithers et al. (2010) and Viveiros de Castro (2004b) suggest, the anthropologist who approaches others from an ontological perspective must leave behind their own theoretical ideas about what nature *is*, what kinship *is*, or even what a relation *is*, since doing so pre-establishes the terrain upon which the native can be encountered and described. In this sense, they advocate an empiricism that keeps the question of theory suspended in order to engage with otherness in a manner that is truly hospitable to other possibilities for thought. To be clear, the point here is not to suspend theory in order to better engage with or understand the interiority of other worlds. On the contrary, the ambition is to suspend the inclination to attach our ideas to the various modalities of otherness we encounter. In this sense, the terminology of suspension should not be equated with a discarding or repudiating theory but just the opposite: the aim is to *make theory vulnerable*, that is, to expose our ideas, and perhaps even ourselves, to the otherness that necessarily arrives from the encounter with difference; an otherness that has the potential to shatter, transmute or develop ideas, opening new dimensions not only for thought but for being in general. This is not because there is no truth (again, the weak fall-back of the epistemological relativist), but because the truth of other worlds is dense and multi-dimensional; it is a truth that is open to revealing itself and effecting our being (our theories, our selves, our modes of existence) in myriad ways. Such a project, yet again, illuminates the unique way these thinkers approach difference. As an example, in Viveiros de Castro (2014), the native's difference is not a set of nounal qualities resident within the native awaiting the anthropologist's inspection. On the contrary, difference is an invitation. The concepts and ideas found in other worlds impinge upon the ethnographer, soliciting us into conversation with other modes of thought. It summons and beseeches us and, in doing so, opens us to other worlds. The ethnographic encounter, in this sense, is not a clash in the traditional epistemological sense – a clash between different world views or discordant cultural representations. On the contrary, it appears more like a crash, as in the wreckage that ensues when always unfolding relational ontologies meet. In this sense, the ethnographic encounter is a siren call. And it is in this smashing against the rocks that new modes of thought can be found.

There are many things that I like about the ontological project in anthropology and I am no doubt sympathetic to its epistemological critique. I also like the way it frames the ethnographic encounter. No longer is the ambition to know and understand another's difference. On the contrary, the ambition is to use this difference; to learn from it by allowing it to infest oneself, subverting not simply the will to know but the question of knowledge itself. And yet, while I very much admire this conception of ethnography and the scene that it sets, there remains in this work a certain tension. To see this, we need to return (once again) to the ethnographic scene: the native talking to the anthropologist about her difference. As already suggested, it is not simply that the native *is* different that grabs the ethnographer's attention. It is that the native speaks about her difference; she speaks about the ideas, values, and modes of thought that characterize her life, and she speaks about them in a manner that names them as distinctive and as belonging to her – her community, traditions, history, and so on.[7] Thus, even as I recognize that the native may speak about her difference using concepts and meanings wholly incomprehensible to my understanding, and even as those languages may crash against my own modes of thinking, transforming and subverting them in the process, there is something about the naming of difference itself or, indeed, the very inclination to name, that introduces a problem: how does one approach, apprehend, or understand these named differences without reintroducing a sense of ownership? How does one accept the statement "these differences belong to me (my community, my tribe, etc.)" without accepting the implication that they are endemic to the community for which the native (ostensibly) speaks? In other words, the problems introduced by ontological anthropology have not substantially changed from the problems that haunted the concept of culture. As Matai Candea (Carrithers et al., 2010) suggests, "The difficult conundrums which dogged the anthropological study of cultural difference do not disappear when we shift to an anthropology of ontological alterity. If anything, the conundrums are sharpened" (p. 179).[8]

In order to illustrate how this tension continues to haunt anthropology, I want to focus on two attempts to resolve it, the first proposed by Mario Blaser and the second by Martin Holbraad. For Blaser (2014), Indigenous ontologies need to be understood not as underlying shared realities but as strategic performances: "My own formulation of ontology constitutes a way of worlding," but in order for this world "to hold it needs to be enacted" (p. 54). In Blaser's Latourian-inflected approach, ontologies are modes of thinking and acting, positively assembled in response to specific situations and problems. To demonstrate, Blaser's sometimes collaborator Marisol de la Cadena (2010) illustrates how the state's refusal to recognize Indigenous spiritual entities (such as Pachamama) instigates Indigenous communities to frame them through the language of "Indigenous rights." In other words, it is precisely because the state cannot see Pachamama through its ontological lens that Indigenous groups appropriate the language of minority rights to make Pachamama visible. Rather than arguing that Pachamama is real (and thus must be *appeased* by the state), they argue that Pachamama is a significant part of Indigenous identity (and thus must be *recognized* by the state). De la Cadena's point is that cultural ontologies are performed in response to contextual situations which (by their nature) are highly political.

For Holbraad, the solution is very different. On the one hand, his understanding of the problem is clear. As Holbraad (Carrithers et al., 2010) states, "Culture, at whatever level of abstraction ... quite properly is said to 'belong' to [people]: 'Nuer' culture, 'Western' culture etc." (p. 185). Thus, Holbraad recognizes the dangers of the proprietary gesture and is seeking to avoid it through his distinct ontological approach. In addition, Holbraad also recognizes that any attempt to name cultural ontologies as different (i.e., to talk about "Nuer ontologies" or "Western ontologies") invariably falls into this same trap. His solution, therefore, is to shift the object of analysis from the people themselves to the ideas they articulate. As he suggests, ontology need not be associated with a particular community. Indeed, it "needn't be territorialized with reference to any geographical coordinates whatsoever" (p. 185). On the contrary, in his view, the aim of ontological anthropology is to use the concepts it finds to explore (1) the various ways the world can be ontologically rendered and (2) how those diverse renderings challenge and subvert our own sense of reality. By focusing on the concepts themselves, Holbraad not only avoids the problem of ownership but he also takes anthropological analysis into a different realm; a place where we can experiment with concepts, exploring how different renderings of the world can meet and be altered in the process. As Holbraad and Pederson (2017) argue, it is through attending "reflexively to the concepts we use" that we can be "prepared experimentally to reconstitute them" (p. 43).

To be clear, I do not think either of these solutions get around the problem of ownership. In terms of the former, Blaser and De la Candena argue for a form of what Spivak (1996) calls strategic essentialism. Similarly, Viveiros de Castro (2014) states that "sometimes it may be pragmatically, that is, politically, vital to describe ontologies as intractable sets of presuppositions" (p. 10). In this framing, there is no problem with naming certain ontologies as belonging to certain groups since those groups (and their ideas) need to be taken seriously. And taking them seriously means recognizing the conceptual and political implications behind different ontological enactments. Indeed, there is a line that Viveiros de Castro (2014) articulates in a number of publications: "Anthropology," he states, "is about the ontological self-determination of people[s]" (p. 18; see also Viveiros de Castro, 2003). While I am sympathetic to the political sensibility of such a statement, it is precisely the equation between ontology and self-determination that I think needs to be brought into question – that is, who exactly is being *determined* by ontology? What difference is being circumscribed and named through such gestures?

The same problem haunts Holbraad's (Carrithers et al., 2010; Holbraad & Pederson, 2017) account. Here, the determination not to associate ontology with certain peoples, places, and communities utterly undermines the notion of ontology. There is a difference between free-floating concepts (even ontological ones) and an ontology. While the former are bricks – to use Deleuze and Guattari's (1994) famous metaphor – the latter represents shared understandings of what is real. The operative term here being *shared*. The nature/culture divide, for example, is not simply a concept. It represents a mode of truth that has fundamentally shaped the outlook, knowledge, understanding, and political actions of Western society. Thus, even if we conceded that ontologies are not necessarily linked to specific people and places, they must be linked to someone – someone who speaks these ideas as indicative of a reality that is theirs. In my reading, Holbraad's ambition to elevate the concept at the expense of people simply closes its ears to the metaphysical problem hiding underneath; that is, in the very act of naming an ontology (as an ontology), we are naming a shared system of knowledge; a system that is owned or possessed by someone. In other words, the proprietary gesture still abides.[9]

To conclude this section, I hope it will be clear in the coming chapters how influential the literature I have been discussing here has been on my thinking. Holbraad, Strathern and, Vivieros de Castro, have been important interlocuters whose ideas permeate this project and I come very close (at various points) to a number of their positions. Yet, I would argue there remains an unacknowledged question lurking in their work about the subject; about the "who" that resides in these different worlds; worlds that are different and where native ontologies reside.[10] How do we theorize the claims we hear emanating from these worlds? If we do not answer this question, we are in danger of yet again naming a self-present subject: a subject who has abiding interior qualities and whose words, practices, and beliefs express an inner ontological world. In the very idea that one can compare one ontological world to another – that one can determine one's ontology as theirs and another as ours – there is an implication that subjectivity is trapped between a remorseless relationality and an abiding difference that makes the native's world its own world. Thus, even as I fully agree that the ontological domains being described do not need to marry or correspond to locatable communities or "cultures," there does seem to be some sense that they abide as domains of thought – as conceptual traditions (systems or institutions of lineage and provenance) *that can be named* – that are (in the worst sense of the term) *cultural*. The very notion of crossing from one domain to another – which Viveiros de Castro's (2004a, 2004b) concept of equivocation implies – means there is a *boundary*. And if there is a boundary, even if it is a highly fluid and transmissible one, then we are obliged not simply to name that boundary but to ask questions about its nature. In other words, the question of "who" exactly claims this boundary must be addressed – not evaded.

A Second Question of Culture

At this point, we have arrived at the central question of this chapter: How do we understand the phenomenon of claiming culture? That is, the phenomenon of claiming certain worlds and the modes of thought, ways of naming, metaphysical cosmologies, and conceptual apparatus therein, as mine? Is it possible to move the question of culture away from considering the worlds that are claimed to the phenomeon of claiming itself? While the previous sections illustrate how the question of culture fails when framed as a question of difference, they also illuminate how the phenomenon of claiming itself (narrating and naming particular modes of life) appears. Underneath the simple recognition of difference there is a claim to difference; the phenomenon of a subject claiming the differences she inhabits as *hers*. And yet how to address this phenomenon (and take it seriously) without naming an abiding interiority – without naming an enduring site (or world) from which this subject who has culture is thought to speak? While (some) ontological anthropologists have pushed away the idea that culture is something that belongs to particular groups, many have not pushed away from the idea that people claim their native ontologies. They do not reject that certain ideas, concepts, and ways of thinking belong to particular peoples and that such modes of belonging are associated with their self-determination. So how to think about this claim in a manner that does not reassert a self-standing, auto-affected subject? If we are going to situate the phenomenon of claiming as the central question of culture, how do we do so in a manner that avoids the pitfalls of traditional anthropology (which explores different species of same subjectivities) while addressing some of the unaddressed questions lingering in various versions of ontological anthropology (which continues to name certain ontologies as "native" even as it affirms their ceaseless becoming). The aim of this section is to illustrate how this project situates the question of ownership and belonging (of claiming ownership and belonging) as its own question, separate from the question of difference.

To do this, however, I turn away from the work of Strathern, Holbraad, Pederson, and Viveiros de Castro and towards the work of Ingold (2000, 2007); another ontological anthropologist who works in a similar, but also distinct, vein. While the authors discussed above are primarily influenced by process ontologies, Ingold is more enmeshed in the phenomenological tradition and draws heavily on Merleau-Ponty (2002), and (to a lesser extent) Heidegger (1971a) and Bourdieu (1990). To be sure, there is more overlap between Ingold (1980) and Viveiros de Castro (2012) than difference. They are both interested in human-nonhuman relations, they both draw upon Latour's (2008) ideas of connectedness, and their ontologies are mutually informed by an active scene of agencies in which natives are embedded. Yet, while Ingold's work is certainly influenced by vitalist ontologies, it is anchored in the language of dwelling over becoming. We can see this most obviously in the way he approaches Indigenous worlds. For Viveiros de Castro (2012), the question of how Indigenous ontologies are constituted holds little interest or meaning (Pedersen, 2012a; Viveiros de Castro, 2004a). Yet for Ingold, this question remains central. One of Ingold's most significant contributions to anthropology (and geography) is his understanding of the role of practical engagement in the constitution of native worlds. Like Merleau-Ponty (2002) and Heidegger (1971a), Ingold (2000) understands practical engagement as the mechanism that allows subjects to inhabit their world. Rather than being a short-hand for tedious passive agency, practical engagement is conceptualized as the silent enabler of being-in-the-world, unlocking the potential of the unfamiliar by incorporating it, as much as it will allow, into the needs of everyday tasks. Thus, being-in-the-world arises through practical, rather than conceptual, engagement. As subjects learn to manipulate both themselves and their world, they develop ways of exploiting its resources and traversing its parameters in a manner that is routine and corporeal rather than reflective or cognitive. In this sense, cultural difference arises from a subject's concrete engagement with (and comportment to) the world it finds; a world that is infinitely deep (offering abundant resources for living and living differently), but also stubbornly material. It is the emphasis on the mechanical, corporeal, and practical ways of living that keeps Ingold's conception of difference anchored in skills, tasks, and habits. Difference, for Ingold, is indicative of the corpus of handed-down manners by which subjects have learned to get on with life (what he terms *wayfinding*) as they skilfully negotiate their environment.

There are two conclusions I draw from Ingold's (2000) framework that I think are particularly salient for this project. First, it seems that the native's world is impenetrable to theory. Obviously, I do not mean that it is impenetrable to being theorized. But for Ingold, the conceptual ontologies discussed by Viveiros de Castro would be a second order phenomenon. What is primary is practical activity: non-contemplative actions, solving immediate problems, using Indigenous resources. The process is impenetrable to theory in the sense that there is no underlying engine or force (no primordial taboos, protein requirements, identity performances, relations of production) guiding worlds towards particular formations. On the contrary, social worlds evolve in relation to the problems that emerge. The shape of the world we are in is the result of an unfolding historical situation; patterns of problems presenting themselves and being resolved using the capacities at-hand. It is precisely this emphasis on practicality that squeezes out other deeper theoretical explanations. While the history and trajectory of any particular social world could transpire differently (indeed infinitely so), the driving force of practicality keeps our understanding of the world permanently grounded in the empirical situation; the practice of beings skilfully getting on with the situation they inherit.

The second, and perhaps more significant, conclusion is that the world Ingold (2000) presents is fundamentally elusive. Different native worlds are not simply impenetrable, they also have infinite potential. For Ingold, this potential is part and parcel of a world constituted not of beings but of a matrix of human and non-human capacities. In this sense, it is as if a native's world is a dynamic field (not dissimilar to Heidegger's notion of the open) that allows various possibilities for comportment and appropriation to be differentially exploited. The world's elusiveness thus resides in the interplay of the dynamic forces available and a being's capacity to skilfully enrol those forces. Ingold illustrates this image effectively with his contemplation of a spider-web on his windowpane. What makes the spider a unique being is its ability to skilfully weave together materials (the window frame) and endemic biological capacities (the hairs on its legs that allow it to climb, corporeal reflexes that allow it to sense vibrations, and of course its capacity to produce webbing). It is the spider's ability to utilize these capacities in a manner that corresponds to its environment that makes life and living (its capacity to catch flies and elude predators) possible. It is also framed by what that environment introduces or takes away (for example, the impact of rain, outdoor lighting, or new predators) and how the spider skilfully negotiates those changes. For Ingold, the spider's being is a composite being because it is predicated upon the relations its life brings together.

There are many things I like about Ingold's work and I am by no means alone in being influenced by his compelling conceptualization of the practical relations between people and environment (e.g., see Carrier, 2003; Gow, 1995; H. Lorimer, 2006; H. Lorimer & Lund, 2008; Lund, 2008; Wylie, 2005). But I am particularly attracted to it because of the two features I described above, features that give Ingold's work a particular existentialist tone. In line with many thinkers working in the ontological vein, Ingold understand worlds as emergent from the relations by which it is expressed – it is impenetrable (it cannot be explained) and elusive (it cannot be predicted or understood). A spider weaves its web around an outdoor fluorescent light to better catch the circling moths. The relation is impenetrable (it simply is) and elusive (it relies upon an unforeseeable potentiality – outdoor lighting – appearing in the spider's world). And while the event can perhaps be described through various terminologies that make the process comprehensible – for example, spider desires, spider intentions, spider skill, spider willing – it cannot be *explained* because the relations do not reveal themselves to explanation. They are practical and thus rely upon the appearance of potentialities that may or may not arise – that occurred via happenstance. This is why, in Ingold's language, the spider is conceptualized as *a composite;* its existence is reliant on a gathering of substances and capacities that defy a full accounting. In this sense, we cannot explain what the spider body *can do* (as Deleuze might say) any more than the spider can. The composition of events, forces, and capacities that constitute the spider's being defies explanation.

And yet, it is precisely this defiance of explanation that invites the question of culture (as a phenomenon of claiming) as a distinctive domain of thought. If we presume that the habitual differences between communities and groups are the result of an inscrutable and elusive process of subjects practically comporting to the world they find, one wonders why subjects simultaneously come to invest in, care for, and identify with those worlds? A world, one could argue, that they had little hand in making. While Ingold's (1976) ethnographic work often seems to suggest that subjects do care about the world they inhabit, there is nothing within his work per se that explains this attachment. Indeed, care simply seems to emerge from the practice of engaged activity (Ingold, 2011). Such an argument seems to subtly lean on Heidegger's own phenomenological conceit, that is, the world is meaningful because it is *ours*. Thus, subjects care for and embrace their world by virtue of the fact that we are engaged in the activities it has bequeathed. Overlooking the tautological nature of this relation, we can argue that such a position once again reinforces the idea that culture is the possession of a self-present subject. It suggests that subjects *have* culture and care for the culture they have because (and only because) it is *theirs*. Such a response does not take on board the full implications of Ingold's (1976) insight. The world we are in is definitively *not ours*. It is elusive and inscrutable: we have neither ownership nor understanding of the world we inhabit. It is something subjects are simply given. To put this another way, the argument here is that *there is* difference in the world. Indeed, it could even be argued that there is essential difference; essential differences that not only separate the ethnographer from the native but also me from you. But those differences are elusive and inscrutable – they are unknowable.[11] We cannot determine the differences we inhabit nor can we determine the differences of others. Our difference defies explanation. It cannot be illuminated. It refuses being accounted for. And yet subjects make claims about their difference. Thus, rather than asking about how native ontologies reflect, represent, perform, or self-determine a native world, the question here is, Why do subjects claim that world? Why do subjects claim that which is essentially not theirs?[12]

The central argument of this chapter has been that cultural theory fails when it is centred on the problem of difference. Following Ingold (2000), I agree that human beings live in habitual worlds; modes of living and existing shaped as we comport to the world we find. But that world is impenetrable and elusive. It cannot be known by us (as ethnographers) nor by the native. Contra Geertz (1983), a guiding presumption of this approach is that local knowledge is not the best knowledge. It is simply knowledge. And while it is knowledge that elicits our attention, our interest and our wonder, this is not because it effectively transcribes or translates a world where others reside. On the contrary, it transcribes and translates a world that others claim; a world that, as Strathern, Viveiros de Castro, and Holbraad point out, is interesting in and of itself. In this framing, native ontologies are native claims. To connect such claims to the world's they name (to see them as indicative of an actual relation of possession) is to fall back on traditional conceptions of culture, the idea that our claims to culture correspond to the distinct compositions that constitute our lives. The aim of this project is to sever this link and in doing so, establish a distinct trajectory of cultural theory; a trajectory that takes the phenomenon of claiming (rather than the ontology it claims) seriously. By this I mean approaching it as an appearance that needs to be explained on its own terms.

In order to do this, however, the question of culture (as claiming) needs to be separated from the question of difference. There are two questions immanent to the question of culture. There is a question about how lives come to be composed (a question that is elusive and inscrutable), and a question about why subjects come to claim those distinct compositions. In addition, while Ingold and others provide compelling conceptions for the former, those frameworks do not provide a foothold for the latter; they do not explain why beings (and perhaps human beings in particular) come to name the world they are in as their own world. The reason for this failure is not, I think, due to any failure in ontological explanations in and of themselves. On the contrary, my point here is that the question of claiming is a fundamentally different question; a question that cannot be thought within the gamut of ontologies currently being explored. There remains, I would argue, a second question of culture: a question that is not about difference but about how and why human beings claim difference. In this sense, this project is predicated upon a key distinction; a distinction that pulls the question of possession apart from the question of ontology. There is a difference between thinking about the native world itself and thinking through the native's fidelity to that world. While I understand the former, following Ingold, Strathern, Viveiros de Castro, and others, as a question of ontology, I understand the latter as a question of culture.

Conclusions

As already suggested, the aim of this chapter (and the two that follow) is to situate the question of culture on a different ground. It began by arguing that theorizing culture as cultural difference is a dead-end. It cannot be theorized again. Not only is this borne out by the historical wreckage of anthropological theory, but also by the unavoidable fact that any attempt to anchor different ways of being in an abiding interiority situates the subject as belonging to her world. The project I am attempting to situate pivots on the proposition that our difference is not our own; that the contours and horizons of our world are utterly incomprehensible and infinitely mysterious. And yet our difference is claimed.

While questioning the origin and nature of the claim has something in common with ontological anthropology, it is also very different. This project begins by accepting certain ontological positions about the nature of cultural difference. In both Ingold (2000) and Viveiros de Castro (2014), there is a conception of native worlds as open-ended systems that unfold and emerge through a vast complex of relations. Whether those relations are understood practically or cognitively does not necessarily matter. What matters is that worlds unfold and become and that the process of becoming has no rhyme or reason. The worlds we are in are inscrutable and elusive. They have no abiding form or shape but, on the contrary, are always already being undone, responding to the various forms of exteriority they encounter. And yet, despite this remorseless system of change, we find human beings making claims about their world. My interest is not in the veracity of such claims but their power – how they are sustained even as the worlds to which they relate often defy them. Indeed, this is precisely the point of such claims. It is the irredeemable absence of certainty – about a world, a self, a culture, a history – that engenders the claim's emergence. As I will go on to suggest in the coming chapters, the origin of the claim resides not in a native's world itself but in the elusive and inscrutable uncertainties that surround her world. In this sense, the orientation of this project is not towards a more radical relationality (as proposed by Pedersen, 2012b) but towards exposing the limits of relationality. The origin of culture, I will argue, resides in domains that are mysterious. But justifying this position, as well as why it engenders something we might recognize as culture, will need to wait until later chapters. The modest aim of this chapter is to illuminate the appearance of the claim itself (as a phenomenon) and to suggest that it cannot be explained via ontological conceptions of native worlds. It is to suggest that the claim opens its own trajectory of thought. There is a question to be asked about why subjects claim their world. And this is the central question of culture.

Chapter Three

Geography's Cultural Legacy

Is not the soul of the other person for us an absolute riddle? Is not man the most impenetrable secret for man? The ideas and the motives of human groups are hidden frequently from an accurate study. They are like a fluid which runs through our fingers.

– Jovan Cvijić, quoted in quoted in R.D. Campbell, "Personality as an Element of Regional Geography"[1]

Two Schools

If the aim of the previous chapter was to illustrate how the question of claiming can be excavated from the tradition of anthropology, this chapter aims to do the same in geography. Though the exercise here is significantly different. In the previous chapter, the emphasis was on establishing the question of culture from the ruins of a defunct tradition, that is, the anthropological tradition of theorizing culture. Geography, however, lacks this explicit tradition. There are no books in geography dedicated to the question of culture and there are precious few extended encyclopaedia entries on the history of culture's theorization from a geographic perspective (cf. Barnett, 2009; Gibson, 2017). In addition, while there is certainly no shortage of cultural geographic concepts – for example, cultural landscape, culture region, cultural ecology – there is no obvious lineage of theorizing what these terms mean or more accurately what makes them cultural (Wylie, 2010). This is not to minimize cultural geography's lasting impact on the discipline nor to suggest that the culture concept has not been an ongoing concern for the field. Rather, it is to suggest that geography's engagement with the culture concept, its theorization of the culture concept, has (with some notable exceptions) tended to be implicit. Landscape as a way of seeing, landscape as text, the diffusion of cultural patterns, regional personalities, signifying systems, all these ideas certainly imply specific renderings about what culture is and how culture works, but they do not constitute the explicit treatment we find in Malinowski (1961), Levi-Strauss (1969), Williams (1971), Geertz (1973), or Clifford (1988). On the contrary, geography's approach to the culture concept is usually tangential. Thus, if there is a canon of cultural theory to be found in the discipline, it is one that requires some excavation. The aim of this chapter is to reconstitute this legacy and, in doing so, illustrate how approaching culture as a claim has a tradition in the discipline.

The chapter is organized around two big points. The first point is that geography not only has a tradition of cultural theory but also a tradition that can be categorized into two main schools. An Anthropogeographical School, which focuses on human-environmental interactions, and a Representational School, which focuses on signification. What distinguishes these schools is not simply the different theoretical concepts they develop, but the fundamentally different ways they approach the question, the different ways they understand what the phenomenon of culture is. In the Anthropogeographical School, the question of culture revolves around the habits, sensibilities, and inclinations of particular communities residing within particular places. This is to say that culture is thought in terms of the ingrained and non-reflective ways communities learn to practically engage with their environment, what Paul Vidal de la Blache calls the *genre de vie* – the habits of life. In the Representational School, the question concerns how communities and groups represent their society and community. Thus, questions about meaning and identity are at the forefront. The emphasis is less on what people do and more on how such doings are thought and represented. The first (and dominant) aim of this chapter is to illustrate these schools by examining some of their chief proponents and the theories of culture they develop. This means looking at the cultural theories of Friedrich Ratzel, Vidal de la Blache (or Vidal), Carl Sauer, the new cultural geographers, and work associated with the cultural turn.

The second big point is that even though these schools represent two different perspectives on what culture is and how culture works, they are both important and necessary approaches. In the final section, I argue that the question being asked by these approaches are questions that need to be asked if we are to understand the phenomenon of culture properly. It argues that culture is both a question of habits (of ingrained manners of doing, practising, and living) and a question of meaning (of investment, care, and representation). It also argues that these are two very different questions and need to be addressed on different terms.

The first question is a question about bodies and their worlds and the second is a question about why subjects come to *invest* in those worlds – that is, why do subjects identify habits of living not simply as habits but as habits that are *theirs*. The aim here is to illustrate how the concept of culture embeds two different and distinct phenomena: a phenomenon of living and a phenomenon of investing, caring, and attaching meaning to those forms of life.

Taken together, these two points lead to a final more implicit point, which is that geography as a discipline is well placed to revive the question of culture in new terms. The bulk of the chapter is dedicated to illustrating geography's history of struggling with the question of culture through these two schools. While the former approach has once again become dominant in geography and anthropology (in particular in the work influenced by new materialism and non-representational theory), the latter approach still matters. The point here is not that geography needs to return to questions of representation at the expense of habit. On the contrary, it is a call to return to questions of representation in a manner that acknowledges habit as a necessary component of understanding what culture is and how culture works. And yet thinking about habits alone only gets us so far. While anthropogeographical theory can help us understand how bodies come to embed certain ways of living, it cannot help us understand why subjects come to claim those forms of life. There is another question (a second question) residing at the heart of process theories of life and living; a question about why subjects care about those ways of living and why they claim them as their own. This is what I call the second question of culture; a question that has been ignored in recent years and needs to be redressed.

Before diving into this argument the chapter needs to illustrate these two schools. It does this by focusing on four key theorists/theoretical movements: Ratzel, Vidal de la Blache, Sauer, and new cultural geography. In each of the following sections, the emphasis is on excavating the latent theoretical architectures concerning what culture is and how culture works. The chapter concludes by illustrating how and why the question of culture needs to be thought through both schools and why geography is well-placed to do so.

Ratzel and the Tradition of Anthropogeography

It would seem uncontroversial to locate the origin of the culture concept in geography with Ratzel's *Anthropogeography*; a text that was as formative for geography as a discipline as it was for sociology and anthropology. Both Boas and Durkheim were enthusiastic readers of Ratzel and in many ways sought to establish their traditions in distinction to his programme for human geography (Speth, 1978). Writing at a time when geography was first and foremost a discipline of landforms, Ratzel sought to establish the role of people in shaping the earth's distinctive regions and places. In his study of the Americas, he emphasized the role of migration, settlement patterns, economy, and technology in shaping the continent. In his study of the Mediterranean, he explored the role of trade, civilization, and hybridization in explaining its unique geopolitical position (Petri, 2016). Both early studies were to form the basis of his anthropogeography; a work whose goal was not only to establish a distinctive domain of human geography but to do so on the basis of establishing the human mechanisms behind geographical differences. To be sure, these mechanisms were indicative of their time, in the sense that they are articulated in the language of natural laws and causal relations (Lossau, 2009). But as Natter (2005) and others (e.g., Livingstone, 1993; Malpas, 2006, 2008; Peake, 2017) argue, one needs to look beneath this language to appreciate the subtlety of Ratzel's endeavour.

Ratzel's theory of culture is predicated on his conception of *volk*, which translates as people or group, and how the *volk* is constituted through its relationship with the earth or soil (*boden*).[2] For Ratzel, *boden* is the community's unique topographical, geological, and biogeological homeland and as such it instils certain characteristics into the *volk*'s nature and temperament. While it is tempting to understand the relationship between *boden* and *volk* in a deterministic manner, it would be more accurate to characterize it as interdependent (Hunter, 1983). Without question the constitution of the *volk* is ontologically linked to the environment. As Ratzel states, *boden* is essentially interlaced with the *volk*'s entire being (see Imort, 2000, p. 60). Yet, as Mercier (1995) also argues, the relationship is more interactionist than it may at first seem. Critically, it would be wrong to understand the relation as a genetic determinism, whereby the environment is thought to directly shape temperament and nature. On the contrary, for Ratzel, the environment worked to shape and constrain the *volk*'s options and capabilities and – in the process– also its personality. As Ratzel states, "There never was a time when man could, without trouble, acquire food, shelter, livelihood, by drawing upon Nature. Nature nowhere brings the food to his mouth, nor roofs his hut adequately over his head." On the contrary, "the various artifices by which [man] manages to exploit what Nature freely gives indicate *a certain development of the faculties*" (Ratzel, 1896, quoted in Mercier,

1995, my emphasis). In this sense, *boden* structures the way the *volk*'s habits, practices, and personality develop and in doing so provides a natural homeland for the *volk*'s distinctive lifestyle. As Lossau (2009) suggests, the *boden* should be thought less as a habitat and more as the *volk*'s natural and most fitting home. It is the environment from which the *volk* emerged – with its unique manners, habits, personality – and from which it continues to draw its sustenance and vitality. It is by adapting to a particular *boden* that the *volk* becomes bonded to that place. As such the *boden* operates as a cultural hearth for the *volk*; a spiritual homeland through which the *volk* is nourished and replenished.

Without question, Ratzel's staging of the relationship between *boden* and *volk* anticipates the blood and soil epistemologies of early twentieth-century eugenics, particularly in the United States, and it is no surprise that Ratzel's work is often accused of being essentialist. Yet, as Natter (2005) and Lossau (2009) argue, it is Ratzel's interest in migration and movement that vividly illustrates the dynamism and openness of his framework. For along with the natural bonds that attach a people to its land, there is also the law of *Lebensraum*, the immanent force that compels a people to expand and grow beyond the confines of their original settlement (W.D. Smith, 1980). For Ratzel, *Lebensraum* is not a mechanistic compulsion but a natural outgrowth of a people successfully thriving within their domain and their need (based upon demands of resources and economy) to expand. *Lebensraum* thus represents a tension inherent to the *boden-volk* relation. As the *volk* become attuned to their *boden* and thus "fitted" to their environment, they also become restless; seeking new resources, developing new forms of trade and economy, and encountering new environments and peoples. What is interesting about *Lebensraum*, however, is not its status as a theory of expansion but how that theory embeds a particular conception of intracultural encounter. While the *volk* is understood as an expression of its unique homeland and its adaptation to the earth, once that *volk* moves and encounters others – other people, places, and landscapes – it is forced to adapt. In Ratzel's earlier work these changes were primarily thought in social Darwinist terms (Petri, 2016). Thus, Ratzel imagined hierarchies of *volk* cultures where different cultural "strengths" compete for dominance. In his later work, however, such hierarchies were disbanded and his work took on a more interactionist tone. In particular, Ratzel discusses contact zones – border areas where different cultures interact and adapt. In this framing, borders do not represent strict demarcations between *volk* and territory but a site of cultural fluidity and dynamism (Cuttitta, 2014). In addition, Ratzel understood these border processes (and the forms of adaptation and change they facilitated) as essential for cultural advancement (Natter, 2005). Cultures became strong not by staying in place but through accumulating change. Thus, even as *boden* provided a spiritual hearth to which the *volk* would always be bonded, progress meant leaving that hearth behind and engaging with otherness. As Ratzel states, "It is an entirely erroneous opinion to believe that a people is stronger in every regard, the more uniform it is. In fact, exactly in those peoples who have achieved the most, multiple races and nationalities are at work together in achieving political and all the more, economic success" (Ratzel, 1906, quoted in Natter, 2005, p. 184).

In sum, it is often presumed that Ratzel's most significant contribution to geographic thought is in political geography and geopolitics. Yet, I would argue that Ratzel is primarily a cultural theorist whose ideas about politics, expansionism, and the state only make sense when they are understood through his conceptualisation of the *volk*. As W.D. Smith (1980) argues, "A state, for [Ratzel], *was simply the result of a particular people's adaptation to an environment*. The form that a state or an entire culture took was therefore shaped by the relationship to *Lebensraum* and the struggle for it" (p. 53, my emphasis). In other words, Ratzel's political geography is an expression of his cultural geography. The state expresses a distinctive *volk* and the dynamics of expansion, migration and adaptation that are a necessary part of the *volk*'s development. The political dynamics that Ratzel outlines in his *Political Geography* are thus grounded in the cultural dynamics he develops in the *Anthropogeography*. Indeed, I would go as far as to argue that Ratzel's political geography can be thought of as the cultural politics of the *volk*; the politics of different group identities embedded in dynamics of migration, expansion, and exchange.

Sadly, however, it was precisely this cultural element – and its insistence on interaction and dynamism – that was lost in the environmental determinist frameworks promoted by Ratzel's most famous students, Ellen Churchill Semple (1968) and Ellsworth Huntington (1934). But it was not lost on one of Ratzel's admirers in France who developed his interactionist mechanisms into one of the most significant movements in the history of geography – an emerging descriptive science of regions.

Vidal de la Blache and régional personnalité

If the previous section used Ratzel's theory of the *volk* to situate an emergent geographic theory of culture, this section does something similar with Vidal de la Blache's concept of *genre de vie*. While both terms have a distinctly anthropological ring, the latter hews more closely to ideas being promoted by Boas and his followers in the early twentieth century, many of whom were eager readers of Vidal (Buttimer, 1971; Speth, 1978). What kept *genre de vie* distinct from Boas's conception of culture was its insistence on the inseparability of land and life (cf. R.C. Powell, 2015). This is to say that any attempt to understand a culture must include an analysis of the milieu in which that community operates and resides. In addition, for Vidal, any understanding of regions must also be an understanding of culture, that is, a question about the kinds of people that live in a place and the ways of life those places embody. The aim of this section – like the last – is to excavate this cultural dimension and illustrate how Vidal's theory of the region is also a theory of culture.

As commentators have suggested, *genre de vie* can be defined as the social structure by which societies both adapt and are adapted to their environment (Mercier, 2009). In Vidal's (1926) words, the *genre de vie* represents a "geographical bond" that unites people to place in a manner where they are utterly interdependent. The environment is what shapes the *genre de vie* but the habits that come out of it also leave their imprint upon the landscape, such that the landscape and the people constitute "a composite," that is, a set of "heterogeneous beings" held together in "mutual vital relationships" (quoted in Buttimer, 1971, p. 45). This interactive, holistic system is the cornerstone of Vidal's thought. Yet to understand the *genre de vie* properly, we need to clarify its theoretical structure. The key to this, I would suggest, lies in properly understanding its vitalism and its materialism.

In terms of its vitalism, K. Archer (1993) illustrates how Vidal's work embeds elements from Lamarck whose theories (along with Darwin's) were in high circulation in France at the time. Specifically, he focuses on the idea that animate processes are imbued with a latent vital force; as Lamarck (1914) put it: "Every body possessing life … is permanently or temporarily animated by a special force, which incessantly stimulates movements in its internal parts and uninterruptedly produces changes of state" (quoted in K. Archer, 1993, p. 505). While Vidal does not directly cite Lamarck, one can certainly see the echoes of this vitalism in his conception of human-environmental interactions. In Vidal, human-environmental relations are driven by mutual instigations that stimulate each other into various kinds of responses. In this manner, human societies and physical environments actively shape each other into organized units. In addition, the process is predisposed towards self-organization and naturally works to integrate wayward energies and processes into its systematizing fold. This is how geographic regions became increasingly unified even as their (social, economic, and political) relations became more complex. Highly influential civilizations (such as empires) arise precisely because they integrate numerous ethnicities, languages, and cultures. Indeed, for Vidal, the regional units formed by human-environmental relations were robust because they were so adaptable and it was only extreme disturbances (such as war and famine) that tended to undermine their coherence.

In terms of materialism, the relations that anchor and sustain the *genre de vie* are those grounded in what Vidal terms *livelihood*. While Vidal understands human and environmental forces as wholly integrated, he also understands them as two very different things. Environmental forces are primary because they establish certain parameters for the possible. But Vidal emphasizes repeatedly that human forces are ultimately more significant as they demonstrate an ability to overcome natural obstacles. In addition, this ingenuity to overcome is driven by livelihood, the need to materially sustain the community. Thus, even as Vidal flatly rejects economic determinism, the method he establishes for defining a *genre de vie* is no doubt informed by a materialist eye. As Buttimer (1971) suggests, the search for livelihood was perhaps the dominant vector of analysis for determining a *genre de vie*: "Livelihood provided the label, the core around which a whole network of physical, social and psychological bonds evolved" (p. 53). Thus, the key indices of analysis focused on "(1) the fundamental lines of material production as related to local natural resources; (2) the dietary patterns in terms of their local availability or commercial cost; [and] (3) the blend of agricultural and nonagricultural activities within the region" (p. 55). The point here is not that material production is a determinant of *genre de vie* but it is the engine. A single milieu with the same ecological-biological conditions could host a range of different *genre de vie*. Thus, there is no necessary relations between certain climatic-ecological conditions and the kinds of habits that transpire there. But those conditions will always be the driver of *genre de vie*. It is the need to generate and sustain a livelihood that stimulates human ingenuity and forces certain habits to take shape.

Given the discussion thus far we can see how the *genre de vie* operates as a theory of regional integration and the development of regional livelihoods, but it is not necessarily clear how it constitutes something we would recognize as culture. To make this leap we need to understand that Vidal (1911) conceptualizes habits of life not simply as systems of labouring, producing and shaping the world but also as habits that instil certain traits: "systematically organized habits," he states, "are reinforced through successive generations ... which leave their mark on the *spirit*" (quoted in Buttimer, 1971, p. 194, my emphasis). Thus, in his discussion of American regional identity, he connects the American character to its wide-open spaces and its use of technology: "United States social life differs from that of Europe in scale, nature, and culture. There is a low density of population and great distances ... over all one finds the triumph of mechanization and technology. The giant transportation patterns and potential for circulation has influenced the American mind: their habits of living reflect this great potential mobility" (Vidal de la Blache, 1902, quoted in Buttimer, 1971, p. 49). Elsewhere he discusses how the development of "higher civilization" in Europe was due to the extremely varied physical environments, giving "the peoples which surmounted them ... profit by the results of a collective experience gained in a variety of environments" (Vidal de la Blache, 1926, p. 18) as well as by the interchange between these groups through trade and migration (see K. Archer, 1993).

While there is no doubt a certain essentialism operating here, the engine of regional personalities is thought to reside in dynamics that are still very much part of our contemporary theoretical landscape. For Vidal, like Deleuze (1994), there is no strong distinction between habits of body and habits of mind. As we corporeally adjust ourselves to the work we do, those actions impose certain rhythms to the way we think and respond. In this sense, Vidal's work can be seen as part of a broader tradition of thinking culture as something that emerges in and through our relations with place. Ingold's (2000) conception of culture as dwelling is perhaps the most recent incarnation of this perspective. Both Vidal and Ratzel laid the cornerstone for this tradition of thought. While Ratzel emphasized the *boden* as primary and introduced cultural variation and change through the process of *Lebensraum*, Vidal puts greater emphasis on the transformative capacities of human subjects and the vital energies of adaptation. In either case, they introduce an approach to the question of culture that positions place (and the ecological and historical forces therein) to be the constitutive element in the emergence of cultural forms. While Boas (1940) would ultimately argue against any significant role for geography, Ratzel and Vidal pioneered a geographic approach to culture that has had a powerful theoretical legacy, even as it has been underappreciated in the discipline. Indeed, even today, with the discipline's unrelenting emphasis on relationality and the vital interactionism between subjects and environment, there is little recognition of the geographical roots of these ideas.[3]

Taken together, the aim of these two sections has been to illustrate the basic theoretical approach and architecture of what I am terming the Anthropogeographical School of culture. While I have focused specifically on the work of Ratzel and Vidal, it is important to emphasize that they do not simply introduce geographic theories of culture but rather they introduce a geographic tradition or approach to culture. Indeed, for all the insistence that regional geography is a descriptive tradition (a characterization promulgated by its advocates as much as by its critics), there is a clear red thread of theory that runs alongside its oft-critiqued "encyclopaedic" approach (Clout, 2003; Matless, 1992b). A thread that begins with Ratzel and Vidal but is developed by numerous regional geographers along the way (Turnock, 1967), including the regional folklore tradition of Fleur in the Aberystwyth School (Jones & Fowler, 2007), the work of Geddes, Fagg, and Herbertson of the Scottish school (Matless, 1992b), and the debates about regional consciousness that characterizes both humanistic geography (Buttimer, 1976; see also Relph, 1976) as well as the more recent work of Aansi Paasi (2003) on regional political identity. Such work no doubt invites further avenues of cultural theory that could and should be excavated. Yet rather than mining this particular dig-site further, I want to turn to a very different tradition of cultural geography; a tradition that brought culture to the forefront of geography, and developed the culture concept in very different terms.

New Cultural Geography and the Representational School

The second school of cultural theory this chapter explores is the Representational School, exemplified by the subfield of cultural geography and its various theoretical evolutions. The historical anchor point for this tradition is normally the work of Carl Sauer (1925) and the practice of landscape interpretation he invented in the 1930s. In a previous version of this chapter (Rose, 2021a), I focused extensively on Sauer's conception of culture and the unique set of influences he drew upon from the anthropology department at the University of California, Berkeley. Yet, while Sauer's influence is formative, if not fundamental, to the constitution of cultural geography in the discipline, his impact on the theorization of culture is transitory rather than indelible. Indeed, as it has been said many times, while Sauer invented cultural geography, his invention was characterized by an approach – a practice that focused on fieldwork and direct observation – rather than a theory. To be clear, I think this point is debatable. Sauer no doubt did have a working theory of culture. Drawing upon his colleague Alfred Kroeber (1917, 1952), he conceptualized culture as a holistic "civilisational style" that patterned the land in particular ways. Yet to suggest Sauer was a cultural theorist would be going too far. As P.L. Wagner and Mikesell (1962) suggests, Sauer was not particularly interested in the inner workings of culture (also see Wylie, 2007). Thus, any contemplation of the mechanisms creating that civilizational style or maintaining its consistency through time and space is not present in his work.[4] In this sense, it could be argued that Sauer's main interest was on "what culture did" – that is, how it inscribed itself on the landscape and how the patterns it left behind could be analysed – rather than on "how culture works."

Yet if the interworkings of culture were not particularly interesting to Sauer, they were a central concern to a new generation of cultural geographers whose work would come to define the Representational School. The first wave of these theorists were the new cultural geographers emerging in the late 1980s. While these scholars retained Sauer's interest in landscape as an expression of culture, what exactly they meant by culture, and thus what they understood the landscape to be representing, was wholly reconfigured. Drawing from the emerging field of Cultural Studies, particularly the work of John Berger (1972a, 1972b) and Raymond Williams (1971, 1977, 1981), these geographers did not approach culture as a totemic entity that holistically imprinted itself onto the world. Rather, culture was a complex set of social and political processes which – over the next twenty years – they would develop in some detail. The discussion begins by exploring some of the key theoretical developments happening in Cultural Studies and then examining their influence on the discipline.

In many ways, the problem of culture set out by the literary theorist Raymond Williams was one that geographers could readily relate to. As Williams (1981) suggests, the culture concept by the late 1970s had ossified into two dominant paradigms. The first was what he calls a "whole way of life" school, an approach subscribed to by many regional geographers as well as by Sauer and his students. Here culture operates as an inner spirit inherited and accrued through history, which situates a distinctive perception of the world and its attributes.[5] This worldview appropriates new information in distinct ways and in accordance with cultural pre-suppositions that keep it consistent with its dominant mindset. It is precisely this holism, this capacity for all aspects of the world to be subservient to a particular (historically bequeathed) cultural frame, which allows it to leave a unique cultural trace in the world. The second paradigm is what Williams calls a "whole social order" approach, which was subscribed to by cultural ecologists and early forms of Marxist and feminist geography. In this framing, culture is grounded in underlying material needs and circumstances. It is rational and universal rather than spiritual and idiosyncratic. Culture does not come upon people through random historical situations. On the contrary, how people organize culturally is dependent on their material circumstances and how they to respond to them.

Williams (2003) is writing from the latter perspective and thus understands culture as something intimately connected to material processes. Yet he is simultaneously critical of the inclination to conceptualize culture as epiphenomenal, that is, as an expression or consequence of a base economic necessity. As he states, while it is "certainly an error to suppose that values or artworks could be adequately studied without reference to the particular society within which they were expressed ... it is equally an error to suppose that the social explanation is determining, or that the values and works are mere by-products" (p. 31). In response, Williams gives culture a far more powerful role. Rather than conceptualizing culture as tethered to the necessities of economy, Williams (1981) understands it as a central force in its development: "Cultural practice' and 'cultural production,'" he states, "are not simply derived from an otherwise constituted social order but are themselves major elements in its constitution" (pp. 12–13). In this framework, culture comes to occupy a mediating third term between economy and society. It is what he calls a "signifying system" through which "a social order is communicated, reproduced, experienced and explored" (p. 13).

The significant theoretical development here pivots on this transformation of culture from a material pattern to a symbolic order; a signifying system that works to communicate normative ideas about society that are pervasive and often unnoticed. Berger (1972b), who explores Williams's (1981) ideas through the cultural practice of European painting, emphasizes how the painting cannot be seen as something that transcends culture – as something seeking universalist ideals – but as a cultural practice embedded in cultural values: "It is usually said that the oil painting in its frame is like an imaginary window open on to the world, we are arguing that if one studies the culture of the European oil painting as a whole ... it's not so much a framed window open to the world as a safe let into the wall" (109). Berger's point is that European painting gives insight into our own secret lives, our own unquestioned presumptions about what is normal, valuable, and in need of being represented. Thus, it is no surprise that a culture obsessed with property and possession is replete with images of landowners and landscapes. While such images no doubt appeal to our aesthetic sensibilities, they also reflect deeply embedded conceptions of beauty and composition and their concomitant association with ownership and order. In addition, because these values are framed as universal, ideal and eternal, they hide their cultural uniqueness as well as how they reinforce the power relations intrinsic to the society they represent.

This conception of culture is no doubt familiar to most geographers today. The idea that culture is a signifying system that mediates and reinforces social ideologies is one that was readily picked up in the discipline on both sides of the Atlantic and came to define an era. Indeed, cultural geography was well-suited to Berger's (1972b) and Williams's (1981) ideas. As a discipline dedicated to exploring the social, cultural, and political forces shaping the material world, it could take Berger's interest in painting (or Williams's interest in literature) and illustrate the implications for our most immediate environments. Thus, Duncan and Duncan (1988) would argue that our everyday streets, parks, and neighbourhoods are not simply reflections of rationale design but they also are indicative of deeply held normative ideas and that an unquestioning, inattentive approach to these landscapes inculcates us "with a set of notions about how the society is organized" (p. 123), which we may be largely unaware of. Similarly, Daniels (1989) argues that the meaningful nature of landscape seduces us into forms of attachment that perpetuate injustice and ingrained social hierarchies. It is not simply that the landscape is dual – both meaningful and powerful. It is duplicitous – the former disguises the latter.

There are no doubt many things to say about this conception of culture as well as the legacies and debates it initiated, many of which have been said elsewhere (see Agnew & Livingstone, 2011; Cloke et al., 1991; Cresswell, 2013; Wylie, 2007). Yet in thinking through its long-term contribution to the culture question, I would suggest that it maintained cultural geography's holistic conception of culture but modified it in two significant ways. First, it was a concept anchored in material relations and, second, it was theorized as a *symbolic* rather than mechanistic system. Cosgrove's (1983) statement that "the production and reproduction of material life is a collective art" (p. 1) captures this dual sentiment. On the one hand, new cultural geographers understood culture as a system embedded in relations of production, that is, in normative social orders designed to reproduce social hierarchies. Thus, productive relations shape the horizon and contours of the group and establish the normative boundaries of collective life. And yet this is not to suggest that the relationship between the material and the symbolic is one of cause and effect. As Cosgrove suggests, maintaining the boundaries of social life is a collective *art*, meaning that symbolic relations occupy a distinct arena of social practice. It is precisely this conception of the symbolic – as something that mediates the material rather than reflects it – that makes this theory of culture innovative. For new cultural geographers, the landscape was not simply an expression of cultural patterning. On the contrary, the landscape is an arena of social struggle; a site where dominant social hierarchies struggle to represent what society is and should be in our most everyday spaces (Schein, 1997). Everyday geography is the place where we, as a culture, struggle to define ourselves.

And yet, the question about the precise role of the material was one that would continue to be an area of debate within the field. To say that culture was a mediating term – something that negotiated the hard realities of materialist relations with the meaningful experiences of everyday life – was still to buy into a base-superstructure argument whereby culture is primarily determined by relations of production. With the arrival of more post-structural notions of power, particularly those introduced by Foucault, as well as an increasing interest in the different kinds of power operative in society (e.g., racism, patriarchy, homophobia), this materialist frame would come to be questioned. This led to a second wave of new cultural geographers that took culture to be a far more dispersed idea rather than a defined social phenomenon.

Second Wave Cultural Geography and the Cultural Turn

It feels wrong and unfair to dedicate what is ultimately a chapter section to this period of cultural geography. Many in the discipline might argue that the work associated with the "cultural turn" (a period comprising most of the 1990s) *is* essentially cultural geography and that any canonization of the culture concept, while perhaps not beginning here, should end here as it is during this period that the concept of culture begins to seep into every corner of the field. Indeed, this is what is meant by the "cultural turn": the discipline (as a whole) turns towards culture. Not only are cultural geographers finding an ever-widening purview of geographic sites to explore (moving from painterly landscapes to ethnic enclaves, sexed and racialized bodies, and suburban lawns), but other areas of geographic thought (e.g., urban geography, economic geography, political geography) are increasingly looking at their objects through a cultural lens. The cultural turn thus represents a high point of cultural geographical work.[6] It marks a period where human geography in its entirety was exploring questions of representation and where the domain of the symbolic was understood to be a key site of social and political struggle.

To be clear, the aim of this section is not to chart the sprawling conceptual impact, or the diverse interests, that characterizes the cultural turn in the discipline. Such a summary is better suited for textbooks, primers, and handbooks, a number of which have done this very well (K. Anderson et al., 2003; Crang, 1998; Horton & Kraftl, 2014; Mitchell, 2000; Thrift & Whatmore, 2004; Wylie, 2007). Rather, my aim is to explore how the cultural turn impacts the concept of culture, and here I make two interrelated points. The first is that the cultural turn is precipitated by the arrival of a distinct trajectory of post-structural thought in the discipline, in particular the thought of Michel Foucault (1972a, 1972b, 1995). While Foucault is by no means the only significant theorist influencing the conceptual vocabulary of geography at the time (the ideas of Pierre Bourdieu, Judith Butler, Edward Said, Stuart Hall, Homhi Bahba, and others are all very much present), his work seems to capture the zeitgeist both in terms of the analytical problems cultural geographers attend to and the tools they use to understand them. The second point is that there is an inverse relationship here between the explosion of interest in culture that takes place and the intellectual coherence of the culture concept itself. In some ways this should not be surprising. As the breadth of intellectual problems that cultural geographers attend to expands, the capacity of an encompassing notion of culture to capture all of them is increasingly diminished. This is the problem I referred to in chapter 1. Culture shifts from being a phenomenon in and of itself to being a context, scene, or domain where power relations take shape through symbolic (rather than strictly material) struggles. My overall point here is that these two issues are intimately related. As the language of cultural geography shifts away from culture as a conceptual "thing" grounded in materialist relations (the position of the first wave of new cultural geographers), to a broader sense of political struggle waged through symbolic forms (the position of the second-wave new cultural geographers), what culture *is* becomes increasingly diffuse. To see this, however, we need to understand something more about the theoretical shifts this period represents.

For over a decade, the conception of culture guiding cultural geographic work was steeped in a Marxist-humanist perspective. As discussed in the previous section, the second-wave new cultural geographers understood culture to be its own unique domain; a system of meaning and textual interplay that reinforced key values in a normative social order. The key to this perspective is the recognition that while the domain of culture was distinct and (to some extent) self-referential, what made it indicative of an identifiable community, society or group was the fact that it was anchored in a system of production. This is why Cosgrove (1985) refers to landscape as a "visual ideology": it was a pictorial scene that obscured the relations of production beneath it, presenting them as a natural part of the everyday world (also see

Duncan & Duncan, 1988). In this framing, the culture concept continues to adhere to a base-superstructure framework – the economic base operates as an anchor for secondary social formations such as culture. While theorists such as Williams (1995), Bourdieu (1990), Althusser (1990), and others (e.g., Barthes, 1977) sought to conceptualize the superstructure as a distinct domain (that bore within it a diversity of social and political inputs), the base always remained, as Althusser famously suggested, "determining in the last instance" (Althusser & Balibar, 1977). Culture, therefore, continued to be explained via its (dialectical) connection to economic practice. What made the landscape cultural was the fact that it expressed a specific community tethered to a distinct socio-economic formation.

During the cultural turn, however, the link connecting the cultural to the economic is severed; and it is a severing that allows for a significant expansion in the purview of cultural geographic work. Critical to this transition was the work of Michel Foucault. To be clear, the impact of Foucault on the discipline (and the social sciences more broadly) goes well beyond his influence on the concept of culture.[7] Yet his theories on the nature of power (and its link to representational practice), radically reshaped the latent base-superstructure dialectic which, heretofore, operated as the underlying engine of cultural coherence. In first wave new cultural geography, the power of representation resided in its capacity to "fool" or "mystify" society in order to ensure the smooth operation of productive relations. As Matless (1992a) suggests, representation was thought of as "derivative of, subordinate to, and on occasion distortive of, an underlying more basic reality" (p. 45). With the rise of Foucault, however, second-wave new cultural geographers come to see representational practices as producing (rather than simply expressing) power. They explored how the positive creation of symbolic systems or representational categories (categories which give subjects various forms of meaning and identity) were a significant means for establishing authority. While first-wave new cultural geographers understood subjects as essentially born free and secondarily oppressed by negative power structures, second-wave new cultural geographers understood subjects as enabled by power. In other words, it is through the everyday practice of operationalizing taken-for-granted rules and categories that subjects define and understand themselves in positive terms. This means that the very things subjects rely on to be subjects (to identify themselves as this or that person) were intrinsically embedded in relations of power. Authority is laced (through and through) into the meanings, representations, and concepts subjects use to articulate themselves as self-defined beings (J. Butler, 1997).

In terms of how this conception of power impacted the culture concept, we can see it operating in two ways. First, culture was no longer seen as something that hid power but something that produced it. Culture, in short, was a power *technique* or what Barnett (2001) terms, a modality of governance: "culture … is understood not in terms of psychological mechanisms of ideology or consent, but rather as involving *the detailed regulation of social activity*" (p. 11; my emphasis). Rather than simply being a means to further power, culture becomes a specific modality of power; a particular way of regulating and ordering society. This leads to the second impact that concerns how cultural geographers analysed culture. While first-wave new cultural geographers focused on unmasking power (revealing the ideology behind the scene), second-wave new cultural geographers explored how authority was forged through representational acts. Thus, rather than revealing a power system already deemed to be there, second-wave new cultural geographers explored how representation was used to engender authority. This, I would argue, was a key transformation. In moving cultural geography away from the study of society to exploring representation as a technique of power itself, culture came to be seen as relevant to any geographic formation. Whether studying urban planning, international development or economic policy, the question of the symbolic was relevant to how power took shape within these domains. Culture, therefore, was no longer an object but a technique. Representation was conceptualized as an important mechanism by which authority was established.

On the one hand, this generalization made the culture concept applicable to a wide range of geographic questions. But, on the other, it led to the dissipation of culture as a substantive idea. Indeed, while the erosion of culture as a thing sounds good, there is something about such statements that rob culture of its essential phenomenality or more accurately, displaces that phenomenality to operations and processes happening elsewhere. This can be seen most obviously in the gradual movement away from the terminology of culture towards other terms, such as *doxa* (Cresswell, 1996), *discourse* (Schein, 1997), or *hegemony* (Kong, 1999; Sharp, 1996).[8] Even though such terminologies continue to link power to representational practices and struggles, the movement away from culture itself (and towards a more amorphous concern with representation and the symbolic) results in culture being both everywhere (because the symbolic is everywhere) and nowhere (because culture itself never comes into view). Culture, in essence, becomes a weak variable, synonymous with power. And the subfield of cultural geography becomes less concerned with culture per se and more concerned with the cultural dimension of authority.[9]

A Second Question of Culture

Thus far, the aim of this chapter has been to explore two distinct approaches to the question of culture in geography. Ratzel and Vidal represent the Anthropogeographical School and the new cultural geographers (in both their first and second waves) represent the Representational School. While both schools understand geography to be a core element in the constitution of culture, they understand what culture is and the role geography plays in forming it in fundamentally different ways. For the Anthropogeographical School, the question of culture is a question about habits, that is, the ingrained everyday practices that take shape within, through, and/or against particular environments. While the precise theoretical relation between habit and environment develops and shifts, the Anthropogeographical School as a whole represents a concern with how culture emerges from human-environmental interactions; from the process of bodies navigating an inherited material situation and the resources and constraints therein. For the Representational School, the question of culture is about meaning. Thus, the central question concerns how landscapes reflect (and in the process reinforce) internal cultural ideas, values, and understandings. The interaction is not between environments and bodies but between environments and minds. Geography is how culture reflects and expresses the internalized meanings and values of a community and in the process normalizes those ideas within material space.

The aim of this final section is to argue that these two schools constitute two sides of the same coin. The question of culture is *both* a question of habits and a question of meaning. While on the surface this may seem like an easy thing to say (they are both right!), a second glance brings the problem into focus. These two schools do not simply represent different approaches to the question of culture, they also represent fundamentally different questions. By this I mean that despite using the same term, they represent different ways of approaching what the phenomenon of culture *is*. For the former, culture is primarily a corporeal, unreflective and lived way of being while for the latter it is a cognitive, self-conscious, and imaginative set of meanings. While some theorists have explored how analysing the former can lead to an understanding of the latter (e.g., B. Anderson, 2014), I would argue that the question of what bodies do is essentially different from the question about how humans understand, represent, and attach meaning to those doings. And I also would argue that both questions need to be sufficiently addressed if the concept of culture is to have any salience or purpose. In this sense, I am arguing that the question of culture embeds two separate questions. There is a question about bodies and there is a question about meaning. And I am arguing that they need to be addressed separately and on their own terms.

To help clarify this dual problem of culture, I draw upon a recent commentary by Ben Anderson (2019) in which he similarly (though implicitly) identifies two essential questions residing within the question of culture. The first question of culture is one that begins from "the proposition that we are involved with the world through all manner of practical (dis)connections *before* we represent the world to ourselves and others (i.e., before some act of cognitive representationalism)" (p. 3, my emphasis). The preposition here is key. For Anderson, the question of culture *begins* with the recognition that bodies emerge in and through a background of material relations engendering various kinds of activities, practices, and habits. The question of bodies (and how they take shape through vital interactive relations) is thus the first question of culture, a question that is essentially anthropogeographical. The second question of culture, for Anderson, explores how those relations are mediated through terminologies and ideas that give them meaning; terminologies such as experience, knowledge, power, or indeed culture. What is interesting about Anderson's commentary is the way he identifies attempts by geographers to address a traditional domain of cultural geographic questioning (questions about meaning, power, representation, and experience) but doing so in a manner that does not rely upon the field's traditional structural anchors. As he suggests, this is work that retains its focus on the processes of connection and disconnection that allow relations – between human, inhuman, non-human, and other-than-human forces – to engender diverse and expanding forms of life. An excellent example he gives is Cockayne's (2018) work on workplace culture in the San Francisco digital media sector. Here the emphasis is not on workplace culture as an actual structuring process – something that produces ingrained patterns of socialised behaviour – but rather as an *idea* (the idea of "workplace culture") that workers subscribe to. Another example is the emerging work on the corporeal experience of precarity. Here, the emphasis is on the experience of living in situations where subjects sense their capacities being diminished and where worry about the future permeates the affective atmosphere of existing. Power is, thus, conceptualized as personal, felt, and intimate, rather than as something external and imposed. In either case, Anderson (2019) illuminates work that addresses traditional geographic territory while retaining its footing in process ontologies that emphasize relationality, affect, and becoming.

To be clear, I agree with Anderson (2019) that the body of work he cites is indeed trying to reconceptualize traditional cultural geographic concerns while remaining faithful to process ontologies. But I am not sure it is doing so successfully. While their questions certainly bleed into traditional cultural geographic domains (particular in relation to questions of experience and power), it is a body of work that remains quite distant from questions of meaning, significance, attachment, and investment. The reason for this, I would venture, is that there is no convincing theoretical mechanism or explanation in the literature – nor would I argue in process theories themselves – for how habitual unreflective relations become meaningful. Thus, I do not see in Cockayne's (2018) work a compelling theory for how or why the idea of workplace culture he discusses becomes something workers become invested in (something they care for or find meaningful). Perhaps his discussion of penalties for non-conformity provides some explanation for why workers care. But caring because you fear retribution and caring because you find something meaningful are different things. To be sure, there are concepts elsewhere that address the move from passive unreflective action to self-aware self-conscious practice – Deleuze's (1994) distinction between active and passive syntheses and R. Williams's (1977) "structure of feeling" being the most obvious. But even here the question is about awareness rather than investment. It is a question of how bodies come to recognize or become aware of their experience, rather than how they come to identify that experience as *theirs*. While I discuss this issue more directly in the next chapter, at this point my aim is to make a similar point that I made in the previous chapter – that is, that questions about living are fundamentally separate from questions about how we claim and care for those forms of life.

This brings us then to the problem at the heart of this chapter. Given that geography has a long tradition of theorizing culture both in terms of habits *and* in terms of meaning, how precisely do we acknowledge and retain the contemporary paradigms exploring the former, while simultaneously developing novel conceptual trajectories for engaging the latter? The response, I propose, is that the question of meaning is a question that must be tackled on its own terms. While the first question is a question about habits, practices, and worlds, about how corporealities emerge and evolve through relations and dis-relations and how those processes rotate into rhythms and assemblages, the second question is about how such assemblages come to be meaningful. And while the two questions are certainly related, there is no mechanism within the former that works to explain the latter – there is no explanation of how or why subjects come to be attached to the corporealities they embody. We cannot presume that subjects invest in the worlds they engender because they are *their* worlds. Indeed, it is precisely the question of "their-ness" – feelings of ownership, possession, and belonging – that is under scrutiny.

There is, in other words, another question here. And it is a question about self-possession. While subjects obviously do not care for all their habits, the phenomenon of identity is one where subjects make choices about which habits are conceptualized as their own. The second question of culture, therefore, is a question that engages this phenomenon: the phenomenon of claiming. This is the practice of identifying one's ways of living as *my* way of living, the inclination to identify the world in which one lives as *my* world, *my* way of life, something that is my *own*. While theories about affect, assemblages, human-non-human relations, and other legacies of the anthropogeographical tradition no doubt explain how forms of life come to be, they do not explain why such forms come to be self-consciously possessed. This second question of culture is what needs to be further developed if the culture concept is to be restored. And geography, I would argue, has the legacy to do so properly.

Conclusions

The aim of this chapter has been twofold. On the one hand, it has been to argue that even though geography does not have an explicit or obvious tradition of cultural theory, it nonetheless has a distinctive approach. The history of geographic thought on culture has gone back and forth between exploring culture as a habitual phenomenon engendered through interactive material relations, to exploring it as a representational project informed by relations of power. The first aim of the chapter, therefore, has been to excavate this tradition. The purpose of this, however, was not purely academic. There is a second aim here. In situating geography as the focal point of these two traditions, I am suggesting that – as a discipline – it is well positioned to keep pushing questions of culture and identity into new territory. We can see how it has done this already by instigating one of the most innovative rethinks of the culture question in almost eighty years. With the rise of non-representational theory and the current ascendency of relational ontologies, geographers have radically reconceptualized the mechanisms by which assemblages of bodies and spaces become manifest, emerging through human non-human relations or what traditional anthropogeographers called human-environmental interactions. And yet, even as this work opened geography (and geographic conceptions of culture) in significant ways, many of geography's traditional questions have been left behind. Questions that also have a long legacy within the discipline.

The second aim of this chapter, therefore, has been to signal some potential directions for developing geography's second question of culture, that is, the question of meaning. This question, as I will demonstrate in the next chapter, is fundamentally different from the question of bodies, habits, and the formations of relational assemblages. While the former is about how certain modes of life and living take shape, the latter is about why subjects come to invest in those forms of life. This, I would argue, is not only a different question but also a separate question. But demonstrating this will need to wait until the next chapter.

Chapter Four

Consciousness as Claiming

I don't think the ego is something bound to psychology … it is bound to the very fact that we have to cope with … strangeness.

– Jean Laplanche, quoted in Caruth, *Listening to Trauma* (2014)

A Question of Composure

I want to open this chapter with a question about composure: What does it mean to have composure? Or perhaps it is better to ask, What does it mean to lose it? To be composed does not mean to put on a mask or construct a persona in the Goffmanesque sense of the term (Goffman, 1959). Rather it suggests a kind of gathering up: We collect ourselves by smoothing out our features, repinning our hair, wiping our brow. We pull ourselves together after we have slipped, after something in our body – with its various needs and ticks – escaped our attention, eluded our control, got the better of us. Composure is always a composition, a bringing together of different elements (background and foreground, colour and texture, grammar and syntax) towards some overall effect. It does not work by allowing some singular immutable "self" to shine through but, on the contrary, is something collected and held in a manner where the arrangement is an obvious part of the construction. Indeed, there is a vigilance at the heart of composure, a monitoring over potentially porous or transgressive borders. While composure may or may not be the result of a self-conscious plan (a pre-imagined concept, idea, or identity that is corporeally actioned), it is no doubt the result of a self-conscious watchfulness, a concern over one's integrity.

In the previous two chapters I argued – through two separate literatures – that the question of culture needs to be divided into two. First, there is the phenomenon of difference, whereby bodies take on different habits and ways of life through various relational encounters. And second, there is the phenomenon of claiming, whereby subjects come to claim those ways of life as their own, that is, as things they invest in, care for, and narrate in and through a language of belonging. In addition, I suggested that the phenomenality of the former cannot explain the phenomenality of the latter. Indeed, the latter is its own question that demands its own approach. The aim of this chapter is to elaborate on this approach more clearly. This means digging into this proprietary gesture – the self-possessive inclination – which I have been attempting to circumscribe as the key dynamic for understanding culture. In other words, it means elaborating what it means to suggest that the second question of culture (the question of claiming) is a question about ownership, that is, about naming an identity that is *mine*. While this topic is explored in earnest in the next section, this chapter begins the process by way of a critique of current trends within the geographical and anthropological literatures. Specifically, it explores those literatures that conceptualize social and collective life in primarily *practical* terms. The argument I present in this chapter is that the emphasis on practical consciousness and practical modes of existence, while interesting and compelling for many reasons, provides an abbreviated conception of human meaning and investment. This insufficiency is illuminated by the problem of composure. On the one hand, composure appears to be something habitual and corporeal; something we do without thinking or reflection. But there is something more going on here. Composure, after all, is usually done in public or in front of the mirror. There is something active and self-conscious about it that yet does not quite rise to the level of a strategic identity performance. For Phillips (1993), composure represents the essence of self-consciousness, not a particular mode of consciousness but rather a primordial hypostatic event. Consciousness arises when we compose our self. When we lose composure we say "we lose our self." And when we regain it, we say we "pulled our self together." Composure, in this framing, is the advent of consciousness itself. It is how subjects become not simply self-aware or self-perceptive, but rather self-possessed. Subjectivity, in short, is the event of being composed; it is the event of establishing one's borders and defining one's sovereignty. We are not simply practical beings, perpetually in relation with what is around us, we are also self-possessed beings, concerned with our integrity and with keeping our self together.

The arguments presented in this chapter operate as a bridge between this section and the next. While it begins the process of outlining a distinctive theoretical approach to human subjectivity (which is the central task of section 2), it does so by pointing out blind spots in current debates about collective life, particularly in geography and anthropology (which is the task of section 1). In this manner, it continues to illuminate the limits of contemporary approaches to questions of meaning and investment, while signalling the overall theoretical orientation of the argument. The chapter is divided into five sections. The next two sections focus on elaborating two literatures that put practical and corporeal activity at the forefront of human existence: the literature on practical consciousness (section 2) and the literature on habit (section 3). The aim here is to illustrate how "the practical" has come to operate as a dominant way of understanding human practice and to illustrate how this work frames consciousness in a particular manner – that is, as a form of self-awareness. To be sure, there is much overlap in these literatures and their separation is somewhat artificial, but they provide different perspectives on how everyday social life is conceptualized as practical, habitual, and non-cognitive. The next section explores the limits of these ontologies. While they are excellent at illustrating how thresholds of sensation and cognition can be crossed through everyday practical engagements, their conception of human consciousness is limited because it is understood as emerging from material encounters. In response, I illuminate the enigmatic dimension which haunts the material world and solicits consciousness in a different key. The final section develops this idea further through the psychoanalytic work of Jean Laplanche (1999). Specifically, it argues that consciousness emerges in tandem with the subject's recognition of itself as abiding within a world whose materiality is elusive, mysterious, and unaccountable. Again, the purpose here is to illustrate how and why consciousness emerges in proprietary terms. While I take the literature on corporeal relations seriously – indeed I agree with much of its content and often lean on many of its arguments – I also see its approach as insufficient, particularly in its capacity to explain meaning and investment. Consciousness in the practice-habit literature is primarily thought in terms of self-awareness. The aim of this chapter is to conceptualize it in terms of self-possession. In doing so, the chapter lays down a cornerstone for thinking culture as a claim.

Practical Consciousness

The first domain of theory I want to explore is the practical engagement literature. There are without question a number of subliteratures I am wrapping together in this catch-all term, for instance the environmental perception literature represented by Ingold (2000, 2011) and Kohn (2013), the more-than-human wing of non-representational theory represented by H. Lorimer (2006), J. Lorimer (2012), Whatmore (2002), and Hinchcliffe (2008), as well as the material engagement movement in archaeology, represented by Renfrew (2004), Knappett (2005), and Malafouris (2013). For simplicity's sake, much of the discussion here revolves around Ingold who operates as a strategic anchor for discussing the significance of the practical, the material, and the corporeal in the constitution of social life. The first half is a review of some of Ingold's key insights before moving on to elaborate these insights and how they inform a particular perspective on questions of consciousness and mind.

To be sure, Ingold's (2000) dwelling perspective is now familiar ground in anthropology, archaeology, geography, and beyond. Familiar to the point that it seems redundant to review it here even as it operates as the key theoretical touchstone for this literature. In brief, Ingold argues that the dominant representationalist paradigm that defined anthropology throughout the 1980s and 1990s was founded on the idea that the cognitive precedes the practical. Thus, what subjects express in their narrative, ritual, and material culture is a projection of an internalized architecture of meaning and significance. In this framing, subjects work first and foremost from mental schemas; from ideas, imaginations, and expectations that are learned from normalizing routines and reinforced through outward expressions. Drawing upon the phenomenological tradition of Heidegger (1996) and Merleau-Ponty (2002), Ingold (2011) argues that social life is not primarily cognitive but practical. Using Heidegger's famous example of the hammer, Ingold reminds us that subjects do not require a mental schema of a hammer to use it. On the contrary, the hammer only makes sense as it is picked up and put to task in a world were hammering projects exist. "The forms people build," Ingold (2000) states, "arise within the current of their involved activity, in the specific relational contexts of their practical engagement with their surroundings" (p. 186). Our engagement with the world, therefore, is primarily practical – we use the tools we find to pursue the projects we have been bequeathed.

Beyond this basic description, I would suggest two further features that characterize this literature. First, it often propounds a distributed conception of agency. In an effort to disinvest agency from the individual subject, agency is understood as a system of potentialities that various beings are differentially poised to exploit. This distribution can be thought both vertically and horizontally. The former is exemplified by Ingold's discussion of an oak tree. Drawing upon Von Uexküll (1957), Ingold (2000) reveals the oak as an entity defined by its various functional tones, providing "shelter and protection for the fox, support for the owl, a thoroughfare for the squirrel, hunting grounds for the ant, egg-laying facilities for the beetle" (p. 176). Here, the world's properties and presence are conjured up as vertical because they are infinitely deep, revealing themselves only in relation to the needs of a being that senses and requires them. In terms of the horizontal distributions, Ingold (2008), Whatmore (2002), and Hinchliffe (2008) draw upon the vitalist ontologies of Deleuze (1994) and Latour (1993) to recognize the array of agencies that come together to allow forms of life to transpire. Thus, rather than approaching a spider as a singular being, it can be seen as a composite: a network of biological capacities and material resources intertwining to facilitate existence. The agency of the spider is distributed horizontally, incorporating a range of potentialities that the spider enrols to make its way in the world (Ingold, 2008).

The second characteristic of this literature is its emphasis on skill. In an effort to move away from cognitive constructs of experience, this work emphasizes practical intelligence, the capacity to know by intuition and experience. Practical intelligence has a long history in anthropology, particularly in the work of Bourdieu (1990). As another follower of Merleau-Ponty (2002), Bourdieu (1990) likens social life to a game of football; a game that requires an intuitive understanding of how to manipulate oneself in relation to other bodies and the constantly changing dynamics of the field. This intuitive know-how is calculating but not contemplative. It involves applying practical strategies to an intimately understood environment. In a similar vein, Ingold (2011) and Malafouris (2013) illustrate how skill is not simply a learned pattern but a calibrated response to a dynamic material context. In Ingold's (2011) analysis of the blacksmith and Malafouris's (2013) discussion of the potter, there is a recognition that the various materials involved in these activities are always shifting and the body is in a constant state of adaptation. Thus, there is a dynamism inherent to practical engagement where the skilled practitioner perpetually modifies her comportment to materials in response to their changing conditions.

As suggested above, the aim of this section is not simply to describe this literature but to illustrate the distinctive way it conceptualizes consciousness. Here, I would draw attention to a final theme in this literature which is the endeavour to erode the division between materiality and mind. In the work of Malafouris (2013) in particular, he argues that the origin of thought in general (i.e., not just self-conscious thought but all thought) resides in things: "human thinking is, first and above all, thinking through, with, and about things, bodies, and others ... Thinking is not something that happens 'inside' brains, bodies, or things; rather, it emerges from contextualized processes that take place 'between' brains, bodies, and things" (pp. 77–8). Malafouris illustrates this point more fully in his discussion of knapping: "in tool making, all formative thinking activity happens where the hand meets the stone. There is little deliberate planning involved ... but there is a great deal of approximation, anticipation, guessing, and thus ambiguity about how the material will behave" (p. 176). Every mental resource, Malafouris argues, grows out of the process of engagement, that is, out of the meeting between the material (and what it gives and denies) and the body (and what it senses and seeks). This observation leads Malafouris (2004) to argue that "*material culture is consubstantial with mind*. The relationship between the world and human cognition is not one of abstract representation or some other form of action at a distance but one of ontological inseparability" (p. 58, original emphasis).

Some of Malafouris's most interesting discussions come from his theory of agency. On the one hand, Malafouris wants to undermine the agency of the knapper in order to locate agency in the moment of encounter (i.e., between skill and stone). But on the other hand, he wants to hold on to what he terms the uniquely human capacity for self-consciousness and ownership. Even as the axe may indeed be the result of multiple agencies, the inclination by the knapper to say "I created this" and to claim it is a result of "his skill" is another order of agency that needs to be explained. His argument, which is similar to positions staked out by Ingold (2012), Renfrew (2004), and (as I will discuss in the next section) Kohn (2103), is that thresholds of consciousness emerge in relation to dissonant conversations between body and world. For example, while a potter may be skilled in shaping clay, the relationship between the potter and the clay is fraught. The potter may intend to build a pitcher but the material may not abide (maybe the clay is too wet). In this situation, self-consciousness emerges from this obstacle, that is, from the material resisting the potter's intentions. Such events remind us of Heidegger's (1996) broken hammer where subjects become self-conscious of their world as their tools for making their way break down. And yet, while I can see how events of breakage and failure may indeed raise the spectre of an "I" within the field of consciousness, it is less clear to me how they engender the second form of agency that Malafouris (2004) describes: agency predicated upon feelings of responsibility and ownership. For Malafouris, questions of temporality are key (and will be discussed further in section 5), but the structure of his temporal theory is too underdeveloped to get much further. It is for this reason that I turn to the literature on habit: a body of work equally interested in the practical and non-cognitive, but with a more developed conception of self-consciousness at its core.

Habitual Consciousness

While the literature on habit shares a number of conceptual touch points with the work on practical engagement, the two are also quite different. To some extent this can be attributed to their distinct theoretical lineages. The practical engagement literature draws primarily from the phenomenological tradition as well as from actor-network theory (Latour, 1993), while the habit literature has a more identifiable vitalist trajectory that includes Ravaisson (2008), Bergson (1959), and Deleuze (1991). These differences are reflected in the work in at least two ways. First, the habit literature is less concerned with decentring the question of agency. As we will see, habit is seen as a particular type of agency that many beings have to differing degrees. In this sense, there is less emphasis on constantly keeping agency divested of any human quality. This leads to the second distinctive feature, which is that the habit literature is simply less vexed when it comes to talking about humans. Because habit is taken to be a feature of all living systems there is no danger of being "humanist" when discussing the habitual nature of human behaviour or even how habit informs human consciousness. It is for this reason that the habit literature, I would argue, offers a more comprehensive understanding of human agency and consciousness even as it affirms human existence as primarily habitual.

If the previous discussion is dominated by anthropologists and archaeologists, this discussion is situated more squarely in geography. Although key figures reside in philosophy (see Carlisle, 2010, 2013; Malabou, 2008) and Cultural Studies (see T. Bennett, 2013; Grosz, 1994), the geographers David Bissell (2012, 2013, 2015) and J.D. Dewsbury (2011, 2015) have been central in elucidating habit's distinctive nature and exploring its implications for thinking spatial-practice. Throughout their work, they illustrate how habit should not be thought of as a passive rehearsal of norms but as an active modality of engagement with a dynamic world. Felix Ravaisson (2008) is perhaps the most significant character informing their work. While Deleuze (1990, 1991) and Bergson (1959) are also significant, Ravaisson's focused treatment has made his treatise a central reference point for thinking about habit as something more than a short-hand for dumb reiteration. To develop this conception of habit properly (albeit briefly), I will focus on two key features that I think characterize Bissell's and Dewsbury's approach. The first is the double law of habit and the second is habit's intensive nature.

The double law of habit refers to the fact that inculcating practices through incessant repetition both refines the body and, in Ravaisson's (2008) terms, weakens passion. As an action becomes second nature – one that can be performed expertly and without contemplation or thought – the body's perception of that action lessons (or weakens) to the extent that the movement is no longer felt or perceived. Obvious examples include active habits like riding a bike or passive experiences like working in extreme heat. In either case, the weakening of perception is not understood as negative. On the contrary, it signals the body's capacity to accommodate and adapt to change. As Bissell (2012) suggests, the double law illuminates the body's essential plasticity, "this is not an elastic body that returns to an originary form but one that becomes modified through repeated practice. Here, habit gathers rather than disperses, holds rather than releases ... The force of habit ... becomes one of specialization, rather than flexibility." Such plasticity allows beings to not simply accommodate to their environment but cultivate resources for further adaptation and change. Thus, it is precisely because the bike rider's perception of the bike has receded she is better able to sense a challenging terrain and recalibrate. In other words, habit extends the bodies capacities rather than dampens or curtails them. As Shapiro (2009) suggests, habit is an abiding resource for adaptation. It marks the ability to receive, prepare, and transform difference.

The second key feature of habit is its intensive nature. For Ravaisson (2008), habit and routine can be found everywhere; for example, in the opening and closing of flowers, the nocturnal and diurnal habits of animals, the metabolization of nutrients, the division of cellular structures, and the vast universe of micro-level processes that allow life to proceed. Thus, habit cannot be thought of only in terms of visible routines. As Bissell (2013) suggest, "habits comprise not just the actual movement or practice that can be empirically seen or experienced, but the potential for that movement to happen. The repetition of a movement generates propensities and tendencies, and these are virtual forces that entrain and carry" (p. 141). A simple example of these virtual forces can be found in the phenomenon of getting out of bed (Bissell, 2012). While this action happens every day and as such can be recognized as a routine, it is by no means (for me at least) a habit nor a highly refined skill. And yet, the actions needed to complete this routine are reliant on a deep hierarchy of pre-perfected capacities consciously pulled together to complete this powerfully willed act. To suggest that habit is intensive, is to acknowledge the layers of dispositional tendencies accreted in any movement that carries the body along. The implication of this characteristic is it allows us to see habit as what Bissell (2015) calls a virtual infrastructure: a dynamic and evolving architecture of intensities developing and modulating in relation to emerging situations, circumstances, and milieu. Habit, therefore, is not simply the routine contraction of a particular muscle mechanism in relation to a particular encounter. Rather, it is a synthesis of diverse dispositions – established through past practice – that can be actualized, adapted, and sometimes eroded in response to present problems and impending futures. It is only by understanding habit virtually that we can see it not as a set of structured routines but as a set of capacities whose actualization is both dependent on and open to the exigencies of circumstance and situation.

The final concern of this section is how this virtual conception of habit bears on the question of consciousness? For Dewsbury (2015), consciousness is conceptualized as an event that emerges in relation to a particular milieu, that is, it arises as certain virtual capacities are actualized. The perspective is similar to the practical engagement literature in that it is the material affordances immanent to the world, and the extent to which they resist or submit to appropriation, which allows consciousness to arise. Even though habit allows for the ascension of corporeal intelligence at the expense of self-conscious willing, if the materials of the encounter resist, deny or undermine the body, dispositions of consciousness and self-awareness will arise. Such situations are most obvious when the body endeavours to learn something new or when it needs to adapt to sudden change, as Dewsbury (2015) suggests, "the sense of self ... comes about as the body becomes aware of the physical effort to do something when encountering the resistance in any sensation-perception of objects" (p. 38). However, it can arise in other instances as well, such as situations of breakdown, when the body's dispositions are stymied by dissolution or fatigue – for example, in jet-lag or mental or emotional trauma (see Bissell, 2015). It can also arise in situations of disruption when the materials to which a body is accustomed are taken away, such as when fighting addiction or in being evicted from home.

While this approach to consciousness is similar to the practical engagement literature – in that it conceptualizes consciousness as something that arises in response to material resistances – the habit literature also understands consciousness as something that emerges and recedes. As Carlisle (2008) notes, Ravaisson (2008) makes no clear distinction between habit, instinct, consciousness, and willing. Indeed, for Ravaisson the difference between these states is one of thresholds, what he calls a matter of degrees: "it is by a succession of imperceptible degrees that inclinations take over from acts of will" (p. 55). Thus, the movement from habit to will is always in flux, always moving in relation to what is in need of address, that is, in relation to the milieu and the demands that milieu makes upon the body and its dispositions. The idea that consciousness emerges and recedes in relation to various obstacles presents us with a conception of consciousness that is essentially *responsive*. In other words, it is something that arises in degrees and in relation to the manner in which it is called. What I like about this notion of consciousness is that it presents the origin of awareness and willing not within the body but within the world itself. Consciousness is a modality of cognition (a habit) that emerges in response to being called or solicited by the world. The aim of the remaining sections is to illuminate other ways that the world calls us and seeing how these calls elicit their own distinctive form of self-awareness.

Mystery within the Material

To briefly summarize, the aim of the last two sections has been to illustrate how theories of non-cognitive practice conceptualize the emergence of consciousness. In both literatures there is the proposition that non-cognitive corporeal activity is the primary mode of human (and perhaps non-human) experience and that self-consciousness is a second order phenomenon arising from some kind of disruption to skilful-habitual absorption. For practical engagement theorists, bodies are composites and consciousness emerges when the relational structures that hold them together are blocked or disjoined. For habit theorists, bodies are a virtual infrastructure, a set of immanent capacities always already available for opening out to the material realities to which they are disclosed. In either case, consciousness is thought to emerge and recede in relation to the pressures, problems, and obstinacies of the world outside, that is, in relation to *exteriority*.

But perhaps this is too simplistic. As Malafouris (2013) suggests, there are at least two ways we can conceptualize the question of consciousness. On the one hand, we can consider the sense of self that emerges through the execution of skill – the potter's awareness that it is her fingers shepherding the pot into existence. On the other, there is the sense of self that emerges through possession – the potter's sense of responsibility for the pot, that is, the fact that it reflects forms of mastery and creativity that are conceptualized as *her own*. The distinction is useful. The practice-habit literature addresses consciousness primarily in terms of the former – consciousness as self-awareness, awareness of one's body, and awareness of oneself as a body. This is a conception that fits its materialist approach to exteriority. But what about the latter sense of consciousness? Consciousness as an event of self-possession, or more accurately, an event of claiming self-possession? This dimension of consciousness is not accounted for by the literature and to understand it we need to widen our conception of exteriority and the different ways it can solicit consciousness. The specific argument here is that exteriority is not just in the world. There is also an enigmatic dimension, a blank spot or silence within the material that we cannot see. And it is this dimension that solicits consciousness as a claim. While Bissell, Dewsbury, Ingold, and Malafouris focus on consciousness as something that emerges through relations with the world, claiming is a mode of consciousness that emerges from recognizing relationality's limits – that is, when relationality is denied, blocked, or otherwise refused (Harrison, 2007a, 2021). Indeed, we have a tendency to view materiality as something tangible, and thus available for appropriation, connection and use, but the argument here is that the materiality of our world is far more elusive. The wind is material. It is in the world and solicits various corporeal effects. But it would be inaccurate to suggest that we are in relation with the wind. While the wind touches us we cannot touch it back. We either enjoy it or bear it, bracing ourselves against the cold or mourning its retreat in the heat. Thus, even as the wind is a present, tangible, and material force, it also withdraws from our reach, unavailable for relation. To illustrate this dimension of materiality further, I want to illuminate its unacknowledged appearance within the practice-habit literature, particularly in its conceptualization of futurity.

In Bissell, Dewsbury, Ingold, and Malafouris the exteriorities that shape a body's doing and thinking are often described in terms of time. Malafouris (2013), in particular, talks about a chrono-architecture of events that, when broken down, may elucidate the sticky problems and forms of resistance that engender consciousness. Similarly, Bissell (2015) and Dewsbury (2015) talk about habit as an openness to the future and the circumstances that the future brings to bear. In either case, the future is framed in terms of the circumstances that we can see; the events that instigate a body into thought. But the future also marks a domain that we cannot see and that disrupts our concern. As Levinas (1987) suggests, the future is something that never properly arrives. It is by definition always yonder, distant, and forever outside what we can experience. In this rendering, the future cannot ever be an event; it is never a meeting of materialities coming together to give rise to something. Indeed, for Levinas (1987), the future is not even time: "in order for this future, which is nobody's and which a human being cannot assume, to become an element of time, it must also enter into a relationship with the present" (p. 79). The future is what is infinitely distant from the present. It is a dimension of time that cannot be addressed or comported to precisely because it stands outside what can be seen or touched; it is not something a body, regardless of its adaptive capacities, can lay hold of. On the contrary, it is precisely "what … lays hold of us" (Levinas 1987, p. 77). Thus, while the body is certainly open to this future, this openness cannot be conceived as an opening to potential. Rather, it is an opening to a dimension of existence that is invisible. It does not give, provide, or allow for relation. On the contrary, it is what takes such conceits away.

To illustrate this conception of exteriority further, I want to discuss Eduardo Kohn's (2013) wonderful book *How Forests Think*, a text that exemplifies the promise and problems of conceptualizing consciousness as a relation with exteriority thought in material terms. Kohn's central argument is that for both humans and animals, relations with exteriority are mediated by signs. These signs are not the higher register signs we associate with human symbolism but the simple indexical relations that constitute an everyday component of our conscious horizons – for example, steam coming off water means heat, the heaviness of air means rain. The world, as Kohn effectively illustrates, is full of such signs; it is laden with signifiers that tell the interpreter (human and non-human) that something is coming. Through his exquisite and intimate descriptions of Avila, a network of small villages located in the rain forests of Ecuador, Kohn elucidates how signs introduce moments of prevarication that engender consciousness; intervals that connect an event to a potential future that requires a decision. Both humans and animals face these intervals and the decisions they make carry significant consequences. The reading of signs, therefore, is the constitutive event in the emergence of an "I" (human or not). Signs, in essence, present beings with problems. They bring them face-to-face with potential futures – with questions that require an answer – and in doing so elicit whatever thinking capacities are available to respond.

One story Kohn (2013) tells takes place during a hunting trip with two informants in the rain forest. When they find a monkey hiding in the canopy, his companion decides to chop down a nearby palm hoping the noise will scare the animal from its roost. Kohn's interest here is how the monkey interprets the noise of the felled palm. It recognizes the falling tree as a sign of danger even as it is unclear about the precise nature of the danger or where it will lead. Thus, the sign of the felled palm is an event that initiates a certain anticipation of the future. It leads the monkey to consider its present situation and its potential outcome. "We humans," Kohn states "are not the only ones who do things for the sake of a future by re-presenting it in the present. All living selves do this in some way or another. Representation, purpose, and future are in the world – and not just in that part of the world that we delimit as human mind" (p. 41). I understand why Kohn suggests that the future is in the world and I agree with him as far as it facilitates a sense of consciousness and awareness in the monkey. However, I would also suggest that his story illustrates how the future is not *just* in the world. For Kohn, the structure of the sign lies in the relation it situates between two presents, the one that is here and the one that we (or the monkey) imagine might come. But the monkey was eventually killed precisely because it did not know what would come. Such mistakes happen throughout Kohn's book: dogs mistaking a jaguar for a deer; humans mistaking master spirits for prey. While it is possible the monkey is incapable of hearing the existential nature of its situation, a human could. And it is precisely this capacity – this distinctive modality of hearing – that allows us to see exteriority in a different light.

To reinforce the point, I draw upon another story in the book. During a bus ride to Avila, Kohn (2013) relates how the bus gets stranded because of a landslide that wiped out the road. Trapped by mudslides in front and behind, Kohn's understanding of the terrain and the carnage he has witnessed by past landslides instigates in him a profound anxiety: "As my constant what-ifs became increasingly distant from the carefree chattering tourists, what at first began as a diffuse sense of unease soon morphed into a sense of profound alienation. This discrepancy between my perception of the world and that of those around me sundered me from the world and those living in it. All I was left with were my own thoughts of future dangers spinning themselves out of control" (pp. 46–7). For Kohn, the sense of alienation and disconnection is interpreted as his mind separating from the material conditions that allow thinking to arise. His anxiety, in other words, is thought unanchored, spinning away from the material situation that is its origin. I, however, have a different reading.

The possible futures that caused Kohn (2013) to feel anxiety were possible, even as they were mysterious. The landslide opened a sign where the future that was seen by most (we need to turn around, we need to stay the night, our trip will be delayed, these things happen) was haunted by other possibilities; futures that could (perhaps) happen; futures that transcended what could be seen. Indeed, what Kohn's encounter with the mud reveals is not a loss of relation but rather a recognition of that with which he had no relation, a dimension that defies relation. The mud and its potential sliding had no relation to Kohn, to the tourists, to his anxieties or their good humour. It stood outside those relations and those relations had no bearing on whether the mud slid or not. Thus while I would agree that material problems do give rise to thought, I do not agree that this is always a matter of relation. The world also precipitates thought *because it eludes relation*. In this situation, Kohn was at the whim not simply of the world but of something else – agencies of mud and gravity that were utterly unaccountable to him. This is a dimension over which he had no purchase and no power. It was not something he could ever lay hold of but, on the contrary, it laid hold of him. Thus, I interpret Kohn's anxiety not as an attunement to a broken relation but to his recognition of relation's absence. It was a demand to attend to a future whose bearing could not appear within the interval of the sign because signs only mark out the space of the possible. In this sense, his anxiety illuminates a dimension of exteriority that lies beyond the material, beyond what he could touch, feel, or see, but that nonetheless appeared to him, if only as a silence or dark shadow of thought.

The point of illuminating this exteriority is to suggest that it also solicits consciousness. But it does so in a manner distinctive from the ways we have discussed thus far. While I agree with Ravaisson (2008) that consciousness arises out of material engagements and I agree that it rises and falls in response to various thresholds of engagement, I also think there is a modality of engagement that is not being recognized in this framework; a modality defined by denial and by non-engagement. And it is this denial, this unmitigable block to relation, that instigates consciousness in a different key. This is a consciousness that is not simply about being self-aware but about being self-possessed. To develop this idea, I draw upon the psycho-analytic work of Jean Laplanche. A theory that illustrates consciousness as an event that is not simply about being self-aware but about being concerned with the self as a self.

The Enigmatic and the Claim

Jean Laplanche was a student of Jacques Lacan (whose work I will discuss further in chapter 8). I draw upon him here because he outlines a conception of the ego as formed through a series of relations, which he terms *enigmatic*. His theory begins with a discussion of trauma. For Laplanche (1999), one of Freud's most ingenious insights concerns how trauma arises not from one event but two. The first event is the one we commonly think of as the traumatic event; an event that overwhelms the subject because it defies their capacity to understand it, for example, a child who is exposed to sexual assault, a natural disaster or any situation where normalcy is radically overturned. In these contexts, the subject does not have the faculties to process what has occurred and the event is repressed. It is removed from conscious thought not because the subject does not want to think about it, but because she fundamentally lacks the cognitive resources to reckon with it effectively. The second event of trauma reminds the subject of the earlier occurrence. Whether this event is violent or trivial, it triggers a reminiscence and, in doing so, releases the original feelings of disequilibrium and vertigo. In this sense, Laplanche remarks that Freud understood the original event as something that sits in the unconscious "like a foreign body which long after its entry must continue to be regarded as an *agens* that is still at work" (Freud quoted in Laplanche, 1999, p. 65). It is what Fletcher (1999) calls a strangeness that sits at the heart of the subject's psyche waiting to be awoken.

Laplanche's (1999) reinterpretation of Freud's conception of trauma is predicated on two manoeuvres. The first is a radical expansion of the initial traumatic episode. While Freud understands this episode as a specific event, Laplanche conceptualizes childhood in general, and infancy in particular, as traumatic. Childhood, according to Laplanche, is the experience of being surrounded, infiltrated, and overwhelmed by that which one does not understand. The adult world, Laplanche explains, is full of affects and sensations that the child has no capacity to cognate, sublimate, or comprehend. While adults have learned (to varying degrees) to cope with the conscious and unconscious signalling sent by other adults, children and infants have no such capacities. They sense their presence but cannot process their coming and going nor the effects and sensations they produce. The messages of adult life are as mysterious as they are ubiquitous. The example Laplanche often draws upon is breastfeeding, a practice which releases a surfeit of conflicted affects, desires, and emotions from the mother, for example, of love, sexuality, anxiety, frustration, and repression. While these affects are not meant for the infant they nonetheless emerge unconsciously, imposing themselves on an infant body who "possesses neither the emotional nor physociological responses which correspond to the ... messages that are proposed to it" (Laplanche quoted in J. Butler 2005, p. 72). In this sense, the first traumatic event is instilled from infancy. The infant's primary experience is that of being overwhelmed by a mysterious otherness or what Laplanche terms the *enigmatic*.

The second manoeuvre concerns subverting the relation between trauma and the ego. In Freud, trauma is something that disrupts a self-standing ego. Thus, the ego is a primary stability and trauma is a destabilizing event. But in Laplanche (1999) this relation is reversed: trauma is the primary event and its occurrence solicits an ego into being. As he suggests in an interview with Cathy Caruth (2014), "the small human being has to cope with ... strangeness. And his way of coping with this strangeness *is to build an ego*" (p. 30). Thus, for Laplanche (1999), the ego is a psychic structure that arises (that is solicited) as a means to cope with the enigmatic. It is built as a means to take hold of oneself as a subject in the face of too much strangeness. In this sense, the ego is precisely that which arises to deal with a primordial fragility. It marks the subject's desire to stake mastery over itself as a traumatized and overwhelmed being.

Laplanche's (1999) conception of the ego, therefore, has two distinctive characteristics. First, it is precipitated by exteriority. Rather than being the primordial psychic structure, the ego emerges in response to enigmatic otherness. It is precisely the experience of "having been given over from the start," that is, of being primordially beholden to an enigmatic other that overcomes and overwhelms that "an 'I' subsequently emerges" (J. Butler 2005, p. 77). Second, the ego is a failed event. Even as it seeks mastery, it does so on what Laplanche (1999) calls a "Copernican stage" (p. 82), a stage where the ego is sustained by what is outside it. Thus, while the ego seeks self-possession, it can never actually attain it since to do so would be to cut itself off from its sustenance. The ego in Laplanche is a paradoxical project: it is built to retrieve something it can never have. Even as the ego dreams of self-possession, the enigmatic experience is always its origin, a latent fragility that both gives rise to the ego and undermines its consolidation and completion.

At this point it is important to acknowledge that Laplanche (1999) conceptualizes the relationship between the enigmatic and the ego as one that transpires in and through human-to-human relations. His conception of alterity is the alterity of the other person, the otherness that inheres in the unconscious – the desires, anxieties, and drives – that are wholly unaccountable even as they have profound effects on ourselves and others. My aim, however, is to illustrate how the enigmatic resides within material relations: in the dynamics of mud, the capriciousness of spirits, and the desires of animals. In these relations we also recognize the work of the enigmatic; an abiding unaccountability showing itself, ghostlike, within and beyond the material problems we face. In this sense, I would suggest that the enigmatic is in the world and one does not need other minds to hear its pull. The question, therefore, is how to account for the enigmatic – and its solicitation of the ego – within the material engagements that characterize practical life.

One way to see this is through the act of touching. In Ingold, Malafouris, Kohn, Dewsbury, and Bissell there is a lot of emphasis on how consciousness arises through relations of touch. It is through practical acts such as pottery or knapping that consciousness emerges and recedes as the world's materials give themselves over to and resist the body. But one thing the literature does not address is how those materials deny touch. While the world no doubt resists touch, there is little acknowledgment of those moments when touch is refused. In Jean-Luc Nancy's discussion of his heart attack in *L'Intrus* (2000), he explores how the experience of feeling his own heart (the feeling of it touching him) allowed him to recognize it as something that was his and simultaneously not his. Rather than his heart being part of him, an intimate component of his being, his heart revealed itself as an intruder, something foreign and utterly outside his dominion. The heart that sustained his existence revealed itself as something that could never be his, something enigmatic in its operations and direction. The point is that in touching the material agencies that engender our existence, we feel their connection and their withdrawal. We recognize them as elements of our existence and as things outside us, as things that have a life of their own, unaccountable to us or the urgency of our reliance. This is how the enigmatic inheres in the world. It is precisely because the material agencies that sustain life are agencies – *that they have their own lives* – that they also reveal themselves to be both mysterious and infinitely distant.

It is for this reason that I understand the material encounters that characterize everyday existence not as syncretic but as traumatic: events that illuminate the world not as a resource for being but as precisely that which undermines such anticipations. While the world appears as gifts and light, it is its unaccountable temperament, the fickle nature of its bounty, that reveals it as enigmatic. The world's materiality is a vision of beneficence and plenty hiding an arcane and unreliable origin. The trauma of material encounters emerges with the experience of this precarity, that is, with recognizing that beings have no recourse over that which the world gives. The world and its provisioning is never ours.

It is only once we see material encounters in such terms that we can also see how consciousness emerges as a claim. While the claim is an inclination to self-centre, it is not auto-affection. On the contrary, it marks a desire to be a self, or more accurately, a desire to *have* a self, to possess oneself, in the face of the enigmatic. But it is a desire that can never be fulfilled. It is a psychic structure built around a fundamental frailty: "an element of irreducible otherness ... the 'oceanic feeling,' which would be precisely, the perception of the enigmatic as such" (Laplanche, 1999, p. 194). The enigmatic, in other words, cannot be denied. It is always there, thwarting the ambition to take ownership over one's existence. It is for this reason that the claim cannot be thought as an affirmation. One cannot affirm what one does not have. On the contrary, the gesture of the claim is similar to that of denial – a refusal to be held hostage to the darkness and silence that robs subjects of ownership over their being. But I have chosen the term *claim* quite deliberately. The claim is a claim in both senses of the term: it is an assertion, a pronouncement of self-identity and ownership, for example, "I claim this title," "we claim this land," or as the run-away slave Sethe reflects in *Beloved*, "bit by bit … she had claimed herself" (Morrison, 1987, p. 94). But the claim is also an allegation. It is an insurance claim, a legal claim or Mark Twain's famous distinction between those who do things and those who claim to do things; announcements that are dubious, questionable, and deserving of scrutiny. The claim to consciousness is always both of these: something that should be taken seriously even if it is questionable.

Conclusions

As already suggested, this chapter operates as something of a bridging chapter. On the one hand, it aims to give a stronger sense of what I mean by the term *claiming*. My hope is that by introducing the concept now, the reader will have an easier time following the justification that will be developed more carefully in the following section. The chapter thus provides a preview of things to come – a conception of subjectivity and consciousness that situates the concept of claiming and points to how it will be used to rethink the question of culture. Yet, at the same time, it does this by engaging with the relational-materialist ontologies that have come to prominence in cultural theory. While my aim in this chapter is not to deny the primacy of relationality, practical consciousness, or corporeal existence (I essentially agree that beings exist through the actualization of virtual infrastructures and I also agree that non-cognitive engagement with these agencies constitutes our dominant mode of being human), it is also to suggest that questions of meaning, culture, and self-possession cannot be properly addressed through these frameworks. It may be true that higher thresholds of awareness emerge in and through corporeal relations, but they also emerge through non-material relations. And it is these latter relations that solicit consciousness as a claim; a claim not simply about being an "I" but about being "me."

As Laplanche suggests in an interview (Caruth, 2014), "An earthquake could be taken in as a message, not just something that is factual, but something that *means something to you* … something where you must ask a question – why this? why did this happen to *me*?" (p. 33, my emphasis). It is in the roar and thunder of the earth's trembling where we here silent questions about me. Not simply about my body or my awareness but about *me*; my life, a life that is mine, and how it can exist in a world whose foundations are so fragile (Carter-White &

Doel, 2022). It is this situation that engenders consciousness as a claim; consciousness as a wide-eyed insomniatic vigilance over our capacity to be – to exist – in the midst of such fantastical, mysterious, and overwhelming agencies.

This chapter has been the first step in outlining a specific conception of subjectivity and the distinctive mode of consciousness it engenders. The next section approaches this issue with more care and a fuller argument. But it sets sail from the same basic principle – that is, that subjectivity constitutes a struggle for sovereignty; a struggle to be a self-standing agent vis-à-vis an enigmatic world that always already undermines such gestures. It is only when we understand subjectivity in such terms, that we can understand how culture comes to be constituted as a refuge. But these arguments await in the next section.

SECTION TWO

Claiming

Figure 2. View of Malay Roads from Pobassoo's by William Westall.

Dreams of Sovereignty

In a short commentary on the image of the desert island in literature, the philosopher Gilles Deleuze (2004) starts with a lesson in geography: "Geographers say there are two kinds of islands," he states, "*continental islands* [which are] … born of disarticulation, erosion, fracture [and] … *Oceanic islands* [which] are originary, essential islands. Some are formed from coral reefs and display a genuine organism. Others emerge from underwater eruptions, bringing to the light of day a movement from the lowest depths" (p. 9, emphases added). In both cases, Deleuze tells us, islands are the result of a profound strife between land and sea, their morphology emerging (tentatively) and falling (drastically) amidst the obsequious and unrelenting demands of pressure, erosion, water, and weather. Whether islands arise from disarticulation or from an eruptive event, they are born from breakage; their calm surface betraying the agitated dynamics of erosion and accretion. But human existence, Deleuze continues, tends to forget this: "humans cannot live … unless they assume that the active struggle between earth and water *is over*" (p. 9, my emphasis). In literature, the desert island often stands for the opposite of anything that might be taken for shaky ground. They are monuments to accomplishment and will; the virile surge of the earth against the hundred dreary pulls of tide and current. To dream of the island, Deleuze states, "is dreaming of starting from scratch, recreating, beginning anew" (p. 10); it is to dream of fortitude and resilience; of Prospero, whose ingenuity and magic domesticates and softens a brutal and harsh land; and of Moreau who civilizes the beasts by way of vivisection and science. Dreaming of islands is a dream of all that human capacity can accomplish. "Humans can live on an island," Deleuze states, "only by forgetting what an island represents" (p. 10). This is the island's unique conceit, its capacity to transform, displace or veil the vulnerability that is the island's condition with a dream of durability and strength; resoluteness and will. "Dreaming of islands is dreaming of pulling away" (p. 10). It is the dream "of being already separate" (p. 10).

In his last published lecture series, Jacques Derrida (2009) similarly picks up this theme of the desert island in his discussion of Robinson Crusoe – the parable of a man who quite literally builds his own world. Relying on his wits and inventiveness, Crusoe must build modernity from scratch: "Everything happens," Derrida states, "as though, on this fictional island, Robinson Crusoe were reinventing sovereignty" (p. 79). Crusoe's story is not simply one of virility and single-minded dedication. It is first and foremost a story of sovereignty. As Michael Jackson (2013b) states, "The world Crusoe creates is entirely *his own*.

Beholden to no one, he is the self-made man, the possessive individualist, a law unto himself, captain and saviour of his own soul" (p. 102, my emphasis). It is this relation between building and owning, between solitude and sovereignty, which is particularly interesting to me. Crusoe's struggle for mastery over his life equates, in Derrida's eyes, to the mastery and possession of the island itself: "the monarchy of a Robinson," as Derrida (2009) puts it, "who commands everything on his island … the sole inhabitant of his world" (p. 20). The island dream is thus a dream about isolation and sturdy self-reliance (of pulling away), and it is a dream about individual capacity, creativity, and competence (of building one's world). But most of all it is a dream of sovereignty; of being one's own man in one's own world; a dream of owning the world one builds, of having a "*Bildung* of the world, this *Weltbild* or *Weltanschauung*" (p. 113) that is one's *own*.

At first glance, questions about islands, separateness, and sovereignty seem remote from questions of culture. While the former suggests ideal geographies built through individual intentionality, the latter concerns modes of commonality; the habits, inclinations, and meaning systems of those sharing a collective life, as well as the mechanisms that allow such notions to be distributed and repeated through time. And yet, as both Deleuze (2004) and Derrida (2009) make clear, dreaming of islands means dreaming of worlds. To be sovereign is not simply to build oneself. It is to build a place that is one's own. As already suggested, the central proposition of this book is that culture is not something subjects ever have. Rather, it is something subjects claim. Thus the central question of culture is not how do subjects represent, perform, express, or otherwise exhibit their culture but why do they build, cultivate, invest in, and claim a world?

The first section addressed this question by separating the question of habits from the question of claiming. The aim of this section, is to develop what I have been calling this second question of culture further. This means exploring where the ambition to claim the world as *my* world comes from. Where does the ambition for sovereignty derive? To answer this question, this section draws upon the concept of dwelling. For both Heidegger and Levinas, dwelling is used to conceptualize and explain how subjects claim their sovereignty. To understand this assertion, we need to understand that for both authors, subjectivity is not primary. In contrast to the long-standing tradition of thinking of the human subject as the origin of thought, language, meaning, sociality, aesthetics, ethics, and so on, these authors place the origin of subjectivity in a set of pre-subjective primordial relationships; a set of backdrop conditions from which subjectivity emerges. What these pre-subjective relationships are and how they form subjectivity will be explored in the coming chapters. Suffice to say for the moment that both authors refuse to think of the subject as *self-standing*. By this they mean that subjects are not the origin of their own abilities. Their capacity to know, think, sense, or experience is not anchored in mechanisms intrinsic to human nature itself. On the contrary, subjectivity (for both thinkers) is conceptualized as an effect of conditions outside us; situations which we had no hand in making and over which we exert little control.

It is for this reason that the dwelling concept emerges in both their work as a means to explain why the self-standing subject appears. Indeed, even though each author destroys the theoretical conditions that position the subject as a self-standing being, they do not deny that subjects often *appear* as such, as if we ourselves were the authors of our own self-hood. Does not the voice that speaks locate its saying in a mind that is its own? Do not the hands that create, identify the creation as emerging from its self? When I address another or consider their work, I am acknowledging that what I hear and see is something that is *theirs*, that belongs to them in some proprietary manner. To be clear, Heidegger and Levinas do not necessarily deny the uniqueness and singularity of our souls. Indeed, for the latter it is a condition of his thought. Rather, what they are asking is whether that singularity is *ours* in a proprietary sense. To approach dwelling in these terms is to approach it as a desire for sovereignty rather than its accomplishment. Dwelling, as I will illustrate here, does not establish sovereignty – it claims it. It does not establish a home – it dreams it. Dwelling is dreaming of islands. A pangeac dream. But it is only ever a claim.

Chapter Five

Heidegger's Dwelling

We attain to dwelling, so it seems, only by means of building.
– Heidegger, "Bulding Dwelling Thinking" (1971a)

Introduction

In the opening line of his oft-cited essay "Building Dwelling Thinking," it is interesting that Heidegger (1971a) states that dwelling is something that must be *attained*.[1] It's interesting because we do not normally think of dwelling in such terms, that is, as an act, objective, or goal. On the contrary, it usually stands for something far more banal, a descriptive term that characterizes the subject's embeddedness in a set of practical activities and everyday social relations. In the work of Ingold (2000), for example, a dwelling perspective is one that prioritizes the context of human-environmental relations from which material forms arise. Thus, rather than approaching landscapes, artefacts, or field patterns as objects that represent a producer's ideas about his world (as well as the norms, values, and power relations therein), he understands them as expressions of unreflective social activities. This characterization of dwelling – which has been adopted in both geography (Cloke & Jones, 2001, 2004; Gregson, 2007; H. Lorimer, 2006; Wylie, 2002) and anthropology (Malafouris, 2013; Radu, 2010; R. Roth, 2009; Tilley, 2004; Zigon, 2014) – is certainly consonant with Heidegger's conceptualization of the term, particularly in *Being and Time* (1962), where dwelling is equated with a certain passivity or somnambulant quality. Because subjects exist in a world inherited from society, everydayness is an unchallenging place; a domain of comfort and belonging where subjects exist in "tranquilized self-assurance" (Heidegger, 1996, p. 176). In Heidegger's later work, this conception of dwelling shifts. Rather than being a site of anesthetized everydayness, Heidegger presents dwelling as a particular modality of caring for what is ours. To dwell, in Heidegger's later essays, is not simply to exist in an unreflective manner. On the contrary, it is to build a place for oneself. In this sense, dwelling is thought to mark a separation or boundary between the world the subject is thrown into and the world that it builds; a place where the subject can be what Heidegger terms an *ownself* – a self whose potential is actualized.

The central aim of this chapter is to illustrate that dwelling needs to be understood as something more than simply the subject's everyday way of caring for the world it is in. For many years both anthropologists and geographers have appropriated Heidegger's concept of dwelling to operate as a stand-in to signal an approach to places, landscapes, and sites that is broadly phenomenological. While there is nothing egregious or wrong-headed about such appropriations, one of its effects has been to emphasize what might be called the comportmental dimensions of dwelling (dwelling as an expression of relations emerging from everyday activity) over its creative and thoughtful ones. The argument I make here is that dwelling is not simply about being amidst networks and relations. It is also about how subject's claim those relations as *their own*. It is this dimension of ownership that I think is often overlooked (or at least diminished) in many interpretations of dwelling.

Perhaps this is for good reason. Heidegger's later work has a didactic quality to it, particularly in its worry about calculative thinking. Thus, the emphasis of his writing is very much about reducing the human role in dwelling and illuminating the potentiality intrinsic to the world itself. My goal here, to be clear, is not to deny the primacy of the world but to emphasize that Heidegger never lets go of the distinctive role that humans play. This is the central aim of the chapter: to illustrate how dwelling always embeds a claim; a claim that seeks to make the world one has been bequeathed one's own world even as that ambition is always already impossible and failed. Even when Heidegger is exhorting humans to guard and preserve the openness of the world, he is exhorting them to build. Building, in other words, is something that must take place in order for dwelling to transpire. While dwelling is no doubt primary (i.e., it is ontological), the relations that dwelling points to must be created. While there is no building without dwelling (to build is to dwell), there is also no dwelling without building. Whether humans build hydro-electric plants which obscure (if not deny) the natural world which bequeaths this power or they build windmills which are attuned to the gifts the earth provides, all dwelling embeds a claim to sovereignty, that is, a claim to make the world one's own (Joronen, 2012).

To make this argument, I do not focus exclusively on excavating the dwelling concept in isolation but instead focus on four key terms, each of which is instrumental for illustrating dwelling as a claim. These four key terms are *being-in-the-world*, the *Augenblick*, *techne*, and the *fourfold*. The first two focus on introducing some of Heidegger's basic propositions and the second two pertain specifically to dwelling. The first section explores some of the founding propositions on which Heidegger's philosophy rests. While this material will be familiar to most with a passing understanding of Heidegger, the aim is to emphasize how his notion of being-in-the-world situates a problem that Heidegger will devote a considerable amount energy to solving. If subjects are wholly and unremittingly in-the-world – if the nature of their being is essentially a modulation of the world from which they emerge – than how do they ever get perspective on that world? How can subjects be anything except that which they have been bequeathed?

Section 2 introduces one of the mechanisms Heidegger uses to resolve this issue. The *Augenblick* is the moment of vision that allows the subject to recognize itself as a being who has emerged in and through a distinctive historical situation. Rather than presuming the world to be natural and necessary, the *Augenblick* reveals to the subject that its world is an historical world that is open to being modified and changed. Part 3 goes into greater detail on how the subject acts (or can act) upon this revelation. Focusing on Heidegger's so-called middle period, it introduces the concept of *techne* to illustrate how subjects creatively appropriate the world to which they are delivered; a world that unfolds its own forces and materials in a manner that provisions subjects with the means to reshape what they have been bequeathed. *Techne* explores how subjects shape the resources and capacities they have inherited into creative modalities of building.

The final section explores an evolution in Heidegger's thought away from *techne* and towards the fourfold a term that characterizes a comportment towards the world that both creates and preserves. This section makes the point that the human endeavour to build (creatively and poetically) is both engendered and overrun by the unfolding of the world itself. The fourfold is Heidegger's effort to dethrone humans (now termed *mortals*) from their central position in his earlier work. Instead of thinking the world through a mortal's own temporal horizon, Heidegger comes to see it as a general unfolding condition; a condition whereby mortal capacities and contributions become markedly (though by no means wholly) diminished. Thus, while the previous section explains how humans have the capacity to build a place that is one's own, this section emphasizes how such sites are both gifted and taken away by the world itself, that is, the unrelenting movement of the fourfold.

Taken together these four aspects of Heidegger's thought explain why I characterize dwelling as claiming. In saying that dwelling is a claim, I am drawing attention to both the active assertion of space facilitated by dwelling as well as the ontological fragility of all such buildings. In terms of the former, terms like *techne* and the fourfold explain how the subject builds; how it endeavours to use the world it finds to express its world in a distinctive manner. However, the concept of the fourfold mitigates against this idea by reminding us that the same world which gives us the resources to build, also undermines all such buildings. The terminology of claiming captures this dual dimension. On the one hand, dwelling is an assertion: a claim in the sense of staking one's ground (he claimed the territory for France). But on the other hand, all such claims are allegations, pronouncements awaiting confirmation (he claimed he was not drunk). To characterize dwelling as a claim is to recognize its buildings as assertions, that is, material claims whose spatiality is wholly precarious. In this rendering, the monuments and landscapes we see are not expressions of who we are (the representational model), nor expressions of our practical activity (the phenomenological model), but are practical expressions of who we claim to be. They are expressions of sovereignty; claims made in the face of an always unfolding world which the subject desires to forge into something that is its own.

Being-in-the-World

The aim of this section is to lay out some of the basic tenets of Heidegger's conception of subjectivity. Like his mentor Edumund Husserl, Heidegger starts by asking basic questions about how to approach that which appears to us in sensory experience. For Husserl, the essence of objects can be gleaned from how they appear. While Heidegger partially agrees with the approach, he critiques it for valorizing abstract contemplation over everyday ways of knowing. Objects are not known, Heidegger argues, through an objective assessment of their abstract qualities. Rather, they are known (as well as loved, feared, and detested) through practical familiarity. In this manner Heidegger replaces the question "How are objects in the world *known*?" with "How do they *exist*?" – that is, "How do they come to take part in our lives and exist alongside our everyday to-ings and fro-ings?" This question does not concern the whatness of objects but rather how they come to appear within our everyday living. In this manner, Heidegger establishes a new basis for thinking human experience. Rather than understanding subjects as separate from the world, contemplating it from a distance, he understands them as intimately and wholly integrated. What appears to subjects appears because they are part of a subject's world, a world that does not stand apart from the subject but defines the subject along with the objects that surround it.

Such a conception of subjectivity can be broken down into two fundamental terms in Heidegger's thought. Dasein is Heidegger's name for human being.[2] This is a being that is defined by its *situation*: a historically bequeathed social world of projects, roles, ambitions, tools, language, and so on, which the subject devotes itself to throughout its life.[3] As humans who are born into a particular culture with a particular history, they find things "there" that give meaning and purpose to their lives; not only norms, rituals, and beliefs but desires and ambitions (to be a doctor or loving wife) and the means towards their fulfilment (go to school, submit to an arranged marriage, embark on a rite of passage). Such pre-established possibilities constitute Heidegger's second term: the world. To be clear, the world that Heidegger describes is not an abstract world that we all share but a particular world; a specific situation, inherited from the past and manifest in a real place. Thus, being-in-the-world means finding oneself in a specific world, with all its inherited projects and ambitions, with all the necessary tools and equipment, pursuing what makes sense. While there are many potential features of this concept of being-in-the-world that could be developed, for the purposes of this chapter (and the book more broadly), I am going to focus on three that expand this preliminary definition and are relevant to the broader discussion.

The first feature is that the world is a world of visibility. To be clear, I am not using this term *visibility* in terms of sight. Rather, I am signalling that the world is not a single world (a single abiding reality) that all human beings can potentially see. On the contrary, there are many different worlds (inhabited by different kinds of subjects) and what appears in one world as a reality – what is visible and true before us – may not be visible in another. One of the metaphors Heidegger uses to make this point is the clearing. The forest is infinite and dark, but being-in-the-world means living in a clearing, an area or domain that has been cleared by those that came before (like woodsman clearing the trees) and where the light provided makes certain things appear. In addition, the things that appear in the clearing (objects, agencies, projects, roles, norms, etc.) do not do so because we have actively learned about them nor because we have become passively familiar with them through everyday use. Rather, they appear because they are expressions of the same system of relations that engender the clearing as a whole. In other words, things are visible in the clearing not because of their attributes as things but because the clearing itself is one where those kinds of things appear. The clearing, therefore, needs to be understood in ontological terms. I know what a hammer *is* because it appears in a world where hammers (rather than sorcery) are required. Thus, to describe the clearing as a domain of visibility is to understand it as a place where all things within it are always already intuitively known. This is not to say there is nothing to learn, but that the things we learn about are already recognized as *real* by the relations that constitute the clearing itself.

This leads to the second characteristic, which is that worlds are understood ontologically. If we understand ontology as the science of establishing what is real, Heidegger wants to take a step back. It is the world, he argues, that make things intelligible. Things have no reality in and of themselves. What makes a chair a chair is not the fact that it expresses some universal conception of chairness. What makes it a chair is that it expresses the relations of the world in which it appears – a world where chairs are required. Thus, while traditional ontologists puzzled over the existence of things by analysing their constituent properties, Heidegger tells us that such analysis will never reveal the truth of their existence. We simply will not know what a hammer is by an analysis of its parts. While such an analysis might effectively describe the hammer's features and form, it cannot tell us why it exists: What is its purpose? Why does it appear in the world at all? As Dreyfus (1991) suggests, "nothing is intelligible to us unless it first shows up as already integrated into our world" (p. 115).

This leads to the final characteristic, which is that existence is understood as practical. While much has been made of this practical element of Heidegger's thought, it is important to emphasize that much of Heidegger's intellectual struggle boils down to breaking free of the problem that practical existence sets. As a relational totality, being-in-the-world presumes a certain correspondence between the subjects we are, the projects we have been bequeathed and the tools we use to pursue them. The hammer fits the project of shed building, the textbook fits the project of accountancy law, and the talisman fits the project of sorcery. Abstract contemplation is unnecessary as long as the correspondence between tool and project remains undisturbed. For Heidegger (1962), most of everyday existence involves engaging with what is intuitively known and familiar and therefore not needing to think about it too much: "In … familiarity Dasein can lose itself in what it encounters within the world and be numbed by it" (p. 71). As we will see in the next section, a large proportion of *Being and Time* is dedicated to thinking about how Dasein can be shaken out of its practical familiarity in order to see the historical (and thus arbitrary) nature of its situation.

The Augenblick

If the aim of the previous section is to illustrate how the subject is fully and wholly immersed in its world, the aim of this section is to illustrate how the subject comes to have perspective, that is, comes to understand itself in terms that go beyond that which the world gives. This is a tricky proposition. If Heidegger argues that subjects can somehow transcend the world they are in, then that would substantially undermine his founding claim. And yet, without this capacity, the subject would only ever be modulations of the world to which they have been bequeathed. Heidegger's solution is to contemplate Dasein's capacity for perspective, that is, its ability to see beyond the projects and equipment it has inherited. It is this capacity that is the focus of this section. This means exploring how Heidegger understands (1) Dasein's capacity to see its situation and (2) the impact of such a recognition. The second question is key here. The event of perspective that the *Augenblick* inaugurates is not a moment of vision only. It is also a moment of action. To see one's situation is to deliver Dasein to a threshold where a decision must be made; a decision that did not exist before the *Augenblick*. This is why the moment of vision is key for Heidegger. It is an event that summons Dasein into a new mode of contemplation and thus a new mode of acting on and with the world.

But let's begin with the event of seeing. As already suggested, the problem being contemplated is Heidegger's own making. On the one hand, subjects are wholly and completely in-the-world, enmeshed, often unthinkingly, with the projects and roles they inherit using the tools and equipment they find. Yet human subjects do more than simply live out the world they are given. Heidegger (1962) identifies a certain restlessness in everyday familiarity, a sense that not everything is as it seems. From where does this restlessness emerge? Heidegger's answer is that the world we are thrown into – the world that we intuitively understand – is both intimately familiar and simultaneously unfamiliar (p. 233). The fact that Dasein fits the world, is not to say that it always feels at home there. This is because the world Dasein finds itself in – the culture it inherits, the values it enshrines, the norms it enforces, and the projects it legitimates – precedes Dasein's existence. The origin of Dasein's being resides in a situation that Dasein had no hand in constituting. The constellation of meanings and practices that define Dasein's being were delivered to it from what Heidegger (1962) terms *the They* world, the public (or masses) who have given Dasein a world with all its various options for being already determined. Dasein is thus condemned to being "lost" in a world, where all "the ... factical potentiality-of-being of Dasein has always already been decided upon" (p. 248). Heidegger calls this condition Dasein's thrownness. By this he means that the world Dasein has been delivered to has been given to it by others. This embeds a sense of existential estrangement in Dasein; an enduring unease that Heidegger terms *Angst*, an underlying nagging that "pursues Dasein constantly" (p. 189), threatening its familiarity and close-ness in a world that is not of its making.

There are three ways that Dasein can respond to this *Angst*: (1) it flees, embedding itself in the world and the projects the world provides ever more purposefully; (2) it attempts to escape by embracing other ways of doing things (other cultural practices, religions, national identities, etc.), attempting to take ownership of itself by enjoining other worlds; or (3) it reconciles itself to its lack of ownership over its being and embraces the world-it-is-in, even as it has no independent purchase on what that world provides. Heidegger calls this final option authenticbeing because it resolutely faces its existential position. While much has been made about Heidegger's distinction between authentic and inauthentic being (for a start see Sheehan, 1981; Wolin, 1992; Zimmerman, 1986), I would argue that the important distinction for Heidegger is between facing or fleeing *Angst* (see Dostol, 1982). For Heidegger, *Angst* is something that must be addressed. Dasein's call of conscience is a harkening in its depths that compels it to look beyond its "They self" and recognize its placement in a world that is not its own: "The call of conscience has the character of summoning Dasein to its ownmost potentiality-of-being-a-self" (Heidegger, 1996, p. 249).[4] In addition, it is this event of facing – of this turning towards one's existential position and reckoning with it – that instigates the *Augenblick*.

The *Augenblick* can best be described as the moment of vision that reveals to Dasein its situation. To understand this moment of vision, we need to go back to our discussion of the world's visibility – specifically, the discussion of the clearing. What is significant here is that this domain of visibility is not firmly circumscribed. It is a region of light and darkness, meaning that the boundary between what is seen and unseen is a matter of shading. Its boundary is permeable and shifting. This means that all those elements of existence that are unseen or less seen can (potentially) be brought into the light (see Zimmerman, 1986). What the *Augenblick* reveals, therefore, is not simply that Dasein is in a world but that it is in a world whose parameters are not fixed and whose potential is undetermined. The *Augenblick* is a momentary light – a blink of the eye that allows Dasein to see its world *as historical*, that is, as a world that moves and changes through time; a world that is delivered to it from a different time and that moves towards an undetermined future.[5] The *Augenblick* opens for Dasein "the time of being and of a world, an opening that becomes the 'abyssal' grounding of beings that are first granted appearance in and through that very opening" (Heidegger, 1962, p. 126).[6]

Before moving on to the next section, I want to make one final (but crucial) point about the *Augenblick*. For Heidegger, the moment of vision is not simply a seeing but a beholding or witnessing. In other words, it is a moment of vision that requires Dasein to respond. As W. McNeill (2006) suggests, the *Augenblick* is more than an event of vision. It is also "an event of freedom," an event that holds Dasein out towards a moment of action. Like any event of witnessing, seeing is an encounter with truth. It is to behold something, which in turn makes us beholden – that is, obliged to act or testify upon what we have witnessed. In addition, the truth that Dasein sees in the *Augenblick* is its own temporal nature. As already suggested, Dasein inherits its world from the past. Thus the world that defines the very parameters of its being is not a natural or necessary one but an historical one, meaning that it is changeable, transformable, and undetermined. This is why the *Augenblick* is an event of freedom. By illuminating Dasein's own historical indeterminacy, it allows Dasein to come to grips with its own agency. This agency did not precede the *Augenblick* but was delivered by the *Augenblick* itself. It is seeing the world as contingent that gives Dasein perspective and delivers it to a sense of freedom; a position where it can see the openness of its situation and make decisions about the world and its place within it.[7] The next section takes a closer look at this moment of action by attending to conceptual developments happening in Heidegger's so-called middle-period, the era between when he took up his rectorship at Freiberg in 1933 and the end of the Second World War.

Techne

The aim of this section is to illustrate how the moment of action is conceptualized after *Being and Time* in terms of *techne*, the force of gathering that characterizes how Dasein participates in the constitution of its world. The importance of *techne*, as I argue here, is that it describes Dasein's participation in the world in terms of creation and craft, or what I will term *building*. In *Being and Time*, Heidegger primarily thought about Dasein's existence in passive terms. Thus, he talked about projects and roles (practices which were primarily inherited) as the means by which Dasein nurtured and cared for its world. The terminology in Heidegger's middle period shifts to a more active and creative register. Building, as it is discussed below, should not be thought primarily in terms of bricks and mortar. Rather, Heidegger uses the term to signal an act of creation. Thus, the world is revealed by Dasein through creative – rather than simply productive – acts. To see this we need to understand how the concept of *techne* arises in Heidegger's thought and demonstrate how it transmutes the moment of action discussed above into something creative and forceful. Now, rather than emphasizing how Dasein cares for the world it finds, Heidegger begins to think of the world as something that must be continually, creatively, and violently remade.[8] I develop the analysis primarily through a reading of Heidegger's *An Introduction to Metaphysics* (1959), a work that is less concerned with Dasein in particular (though Dasein still figures prominently), and more concerned with the history of being, or more accurately, the history of the way being has been conceived (see Elden, 2001).[9] It should be noted at the outset that this reading of *techne* is predicated on a very singular text that Heidegger clearly turns away from in his later work. This is to say, that while I think my reading is accurate, I do not think it represents Heidegger's overall view on *techne*, even as some of its inflections no doubt carry-on.

The central preoccupation of *An Introduction to Metaphysics* is excavating pre-Socratic conceptions of being, which, Heidegger argues, have been corrupted by Western metaphysics. Specifically, he focuses on a certain dynamic of being in the pre-Socratic world, a movement precipitated by the tension of *physis* and *logos* (Heidegger, 1959, 1977a, 1998).[10] For Heidegger, *physis* (often translated as nature) is the origin of being – its first movement. In his reading of Aristotle, he proclaims that all beings are in movement and exist as long as they have movement: "[Beings] *are* beings only insofar as they have their essential abode and ontological footing in movedness" (Heidegger, 1998, p. 190). Yet *physis* also describes a particular kind of movement, a movement that tends towards a specific direction or orientation. From his reading of Heraclitus, Heidegger (1959) describes this movement as "self-blossoming emergence," the movement of coming forth and appearing: "*Physis* as emergence can be observed everywhere, e.g. in celestial phenomena (the rising of the sun), in the rolling of the sea, in the growth of plants, in the coming forth of man and animal from the womb ... *physis* is being itself, by virtue of what essents become and remain observable" (p. 14). The description of being as that which bursts forth into appearance can also be understood within the dynamics of visibility and darkness (revealing and concealing) that characterize being-in-the-world. In *Being and Time*, Heidegger (1962) makes clear that the world that appears to Dasein is the one revealed through its own distinct modality of being-in-the-world (for any given world that Dasein is in, certain things or aspects of things appear while others remain hidden). *Physis*, however, attributes revelation and appearance to things themselves. In other words, it is not just Dasein that determines whether something appears. Rather, *physis* names a more generalized force that can move things out of darkness; an emergent aspect of being that brings things before our eyes: the startling revelation of a landscape as we crest a hill, the sound of a rushing brook in the distance, a fallow field blossoming forth with flowers. *Physis* is the livingness of living nature itself, an elemental flourishing that goes beyond the capacities of Dasein: "The process of a-rising, of emerging from the hidden, whereby the hidden is first made to stand … [it is] the power to emerge and endure" (Heidegger, 1959, p. 15).

The second movement of being that Heidegger explores is *logos*. Similar to his approach to *physis*, Heidegger excavates *logos* from the history of Western metaphysics where it traditionally languished as logic, "the science of thinking" (Heidegger, 1959, p. 119). Heidegger reinterprets *logos* as *Lesen*, the German word for bringing together, relating or gathering. If *physis* describes the earth's originary emergence, thinking (as *logos*) is the movement that gathers – that which brings what emerges together. *Physis* is the untameable energy, the power that bursts forth and *logos* is the force of gathering, association and collectedness. Critically, however, although *physis* and *logos* constitute a tension, they cannot be conceived as an opposition. On the contrary, he argues, they constitute a unity. While *physis* is being as bursting forth, *logos* gathers –it hears, collects, catches up or, as Heidegger (1959) puts it, *apprehends*. These are two sides of the same coin. As a bursting forth *physis* requires *logos*, the force that pulls together that which comes forward. The latter is predicated on the former, *physis* and *logos* are a unity even as they tend apart: "Being [as *physis*] means: to stand in the light, to appear, to enter into unconcealment. Where this happens … apprehension prevails and happens with it; the two belong together. Apprehension is the receptive bringing-to-stand of the intrinsically permanent that manifests itself" (p. 139). *Physis* comes forward and *logos* allows what comes forward to stand; "*physis=logos*," Heidegger states, "being, overpowering appearing, necessitates the gathering which pervades and grounds" (p. 175).

But how do these concepts relate to Dasein? Unlike in *Being and Time*, Heidegger (1959) does not argue that the misapprehension of being is immanent to Dasein's being-in-the-world (defined as it is by the They world which precludes Dasein from seeing its being properly), but rather, is a result of history itself. Since Aristotle, and the equation of the *logos* with logic, being has been reduced to that which shows (that which is present), and the potentiality of being has been lost. In recovering a conception of Dasein's being in the inter-play of *physis* and *logos*, Heidegger endeavours to illustrate how Dasein is itself an effect or expression of the *logos*. In so far as Dasein's being *is* being-in-the-world, it expresses the worldly context in which it is gathered.

This leads to Heidegger's discussion of *techne*, the expression of the *logos* that belongs to Dasein. While *techne* is traditionally understood in terms of skill – the artist's craft, the poet's wit, the philosopher's cleverness – for Heidegger it is an elaboration of Dasein's capacity to act. *Techne* is not the skill of a particular Dasein (e.g., the skill of the artist), but a capacity delivered to Dasein from the world itself via its dynamic unfolding (*physis*) and its ability to see into its own legacy by virtue of the *Augenblick*. In this sense, *techne* characterizes the unique manner in which Dasein participates in the unfolding of its own temporal horizon; a mode of participation that in apprehending its inheritance, Dasein can effectively gather it, shape it, and release it towards the future. "Whoever builds a house or a ship," Heidegger (1977a) tells us, "reveals what is to be brought forth ... *in advance* ... what is decisive in *techne* does not lie at all in making and manipulating ... but rather in the aforementioned revealing" (p. 13, emphasis in original; see also Heidegger, 1971c). *Techne* is thus Dasein's capacity to creatively reveal by seeing in advance *what is possible* given one's heritage. It marks Dasein's ability to capitalize on the freedom provided by the *Augenblick* and the threshold of action it precipitates.

The point of this discussion is to illustrate how *techne* situates building as Dasein's unique way of being-in-the-world. While the force of gathering happens everywhere (e.g., the basin gathers the river that gathers the herd), *techne* signals Dasein's distinctive capacity to gather in a manner that has perspective (a capacity once again provided by the *Augenblick*). Thus, it is an ability that belongs to Dasein and Dasein only. More importantly, it is a capacity that reveals itself through *building*. If we understand Dasein as a being who is uniquely free to act (unlike animals) and if we understand such actions as grounded in Dasein's capacity to creatively gather (to produce by seeing in advance the possibilities inherent to its own illuminated potential), then such revelations need to be understood in terms of real things built. By this I no doubt mean homes, monuments, community gardens, and amusement parks but also paintings, poems, traditional dances, and folk craft. All such creations are buildings since they execute a world through the material operationalization of its potential (see Elden 2001). In addition, while Heidegger clearly favours some form of *techne* over others (he disparages modern technology at the expense of traditional crafts), following Harman (2002), I would suggest it is important not to be distracted by these distinctions. The key point of *techne* is that it builds with what it finds; a form of practice that Heidegger describes as *violence*. *Techne* is not a tired gathering or a passive pushing forward of possibilities but is a force that tears, digs and contends. It is through violence that the well-trod crust of the everyday is shattered; not a violence of devastation but of renewal or as Elden (2001) suggests, of "unbuilding rather than obliteration" (p. 87).[11] When Dasein digs into its heritage and gathers that which it finds, it takes care of its world in all its depth. In other words, it takes care of its world in all its potentiality. *Techne* allows humans to discover their destiny: "We do not learn who man is by learned definitions; we learn it only when man contends with the essent, striving to bring it into being, i.e., into limit and form, that is to say when he projects something new" (Heidegger, 1959, p. 145). In doing so, Dasein must be "the violent one, who tending toward the strange in the sense of the overpower (i.e. *physis*), surpasses the limit of the familiar" (Heidegger, 1959, p. 151).

The example that Heidegger returns to throughout his later work is poetry. While the everyday world of language is filled with idle talk and senseless chatter, poetry digs into the heritage of language and retrieves from it something original: "Poetry like thought must engender a breaking – a violence – if it is to express something outside historical Dasein" (Heidegger, 1959, p. 26). While Dasein is always in language (i.e., it is always already in the language it has been given), it is not at home in language. "The one returning home," Heidegger (2000) states, "has not yet reached his homeland simply by arriving there" (p. 34). In Heidegger's (1959) discussion of Holderlin's poem *Homecoming*, he suggests to be at home in-the-world and in-language, requires more than mere existence, it requires breaking into "what has been granted, and yet is still denied ... the still withheld essence of [the] homeland" (p. 33). What is granted is language itself, but what is denied is the full potentiality of language (its essence). To break into this potential requires violence to be "the liberation of language from grammar into a more original essential framework" (Heidegger, 1978a, p. 194). It is through such violence that Dasein can build a home for itself. A home that is not a comfortable corner but a place where one is *a stranger*; a world that allows the potentiality of language to emerge and reveal itself as a strange and shattering light.

The Fourfold

The central aim of this chapter thus far has been to make the point that building is Dasein's unique mode of being-in-the-world. Building, as *techne*, is how subjects create a home for themselves. And yet, it is at this point that the argument hits two important snags. The first is that such a characterization does not fully match with the central argument of this chapter. The argument I put forward is not simply that dwelling describes how subjects build. It describes how they build a place that is their own. It is this sense of sovereignty that I want to emphasize here. Dwelling is how subjects build a place for themselves in the midst of being. This brings us to the second snag, which is that Heidegger's conception of building profoundly changes in his later work. Most significantly the concept of *techne* is dropped in favour of the concept of *poeisis;* a concept that deemphasizes Dasein's unique capacity to creatively build in order to put even further emphasis on forces outside Dasein and indeed outside Dasein's world. Thus, while the discussion of *techne* in the previous section understood building in relation to *physis,* Heidegger expands this dualistic relationship through his conception of the fourfold. In the same way that the dynamics of *physis* and *logos* expanded Heidegger's notion of world (away from a Dasein-centric perspective), the fourfold pushes this idea even further.

In Heidegger's (1971e) essay "The Thing" he describes the fourfold thus: "earth and sky, divinities and mortals – being at one with one another of their own accord – belong together" (p. 179). As ambiguous as the concept of the fourfold is, it can in fact be encapsulated through this simple phrasing: "being at one with one another … belonging together." As a number of commentators argue, the fourfold cannot be understood by trying to define its individual elements. What matters, as Harman (2009) suggests, is their "mirror-play"; the way they unfold together in a unity, each element inextricably bound to the others (also see Haar, 1993; Harman, 2002; Zimmerman, 1986). Thus, while any discussion (including the following) must begin with an explanation of the fourfold's elements (earth, sky, gods, and mortals), it is important to not get distracted by the elements themselves and focus instead on their conceptualization *as a four*.

In "The Question Concerning Technology" (Heidegger, 1977a), the earth is described as the matter (*hyle*) of things, that is, the wood of the cross, the clay of the pot, the sand of the glass. In "Origin of the Work of Art" (Heidegger, 1971c), it is the endless resources buried and preserved beneath the earth's surface; the material potential of the world that withdraws from human mastery and control. The sky on the other hand, opposes the earth. It is the dimension that brings materials into "the light and dusk of day" (Heidegger, 1971a), opening the earth up and setting it apart. In the language of Malpas (2006) and Haar (1993), the sky "creates space" for things to appear, providing an opening for beings to be what they are. The gods are described as divine messengers (Heidegger, 1971d), beings that give voice to the destiny of the community by presenting a correlative by which mortal actions can be judged (Young, 2000). Like the endless materials deposited in the earth, the gods stand for the eternal presence of a past bearing on the present. Mortals, on the other hand, are beings that face death. While death plays a central role in *Being and Time* (Heidegger, 1962), it is primarily framed in negative terms – the finality of death that throws Dasein back on itself (see Young, 2000, 2006). Here however, death stands for Dasein's mortality. It is not simply that Dasein will die, but that Dasein *is not* a god. It is a mortal being inhabiting a mortal (and thus changeable) world (see Malpas 2006).

To think of these elements as individual agencies is, as previously suggested, not very productive. But thinking of them as two distinct (but related) axes *is* useful. As Young (2006) and Malpas (2006) suggest (though in different ways), the fourfold represents the coming together of two distinct expressions of a relation that permeate Heidegger's thought, that is, the relation between present-to-hand and ready-to-hand, or more simply, the relation between the object that appears, and the background contexture of relations that constitute its appearance (Harman, 2002). In the earth-sky axis, we can see the relation of *physis* and *logos* (the earth's potential actualized through its own self-blossoming emergence), and in the gods-mortal axis we can see the relation of *techne* (Dasein's actualization of the potential in its heritage).[12] This is why the fourfold can be seen as an expansion of the relations already discussed. The power of the fourfold comes from understanding all four. It is the mirror-play of all the elements (earth, sky, gods, and mortals) that we understand the extent to which an object reflects the world's relational totality. Heidegger's (1971a) example of the bridge remains, I believe, one of the best articulations of this situation:

> The bridge ... gathers the earth as landscape around the stream ... [it] is ready for the sky's weather and its fickle nature ... the bridge lets the stream run its course and at the same time grants their way to mortals ... [it] leads from the precincts of the castle to the cathedral square ... brings wagons and horse teams, to the surrounding villages ... gives to the harvest wagon its passage from the field into the village and carries the lumber cart. (p. 152)

When seen in light of the fourfold, the bridge draws together a "peasant world" *in its inter-relational totality*: the landscape of meadows, dale, and streams; the stone of the arches and the mud that binds them; the bequeathed knowledge and traditions of bridge-making; the agrarian economy of crops, tools, and market towns; the horses and wagons; the floods and rains that the bridge must withstand; the roads and routes of overland trade that rise and fall. These elements encapsulate the peasant's world as *a peasant world*, that is, as a world that the peasant is not simply *in* but is himself expressed through as *a peasant*. It is through this mirror-play of the fourfold (the gathering of landscape, economy, animals, weather) that things in the world (bridges, wagons, horses, grains, peasants, market towns) appear.

The significance of this formulation is that it further reduces the privileged position of Dasein. Indeed, while the aim of the last section was to illustrate the role physis in logos play in Dasein's building, the aim of this section is to illustrate how Heidegger pushes this diminution of Dasein and Dasein's capacities even further. No longer is Dasein the agent of actualization violently bringing certain kinds of building to light. Just the opposite, the bridge expresses human embeddedness in and among forces and materials outside them. The material of the bridge (wood, stone, sand, and clay) is thought of as being delivered to Dasein in the same way that Dasein is thought of as being delivered to the bridge as well as the bridge's necessity given other exigencies already operative in the world. In other words, there is no separation – and thus no temporal order – between Dasein's (now rendered as mortal) making and the bridge's appearing: "to say that mortals *are* is to say that *in dwelling* they persist ... by virtue of their stay among things" (Heidegger, 1971a, pp. 156–7, original emphasis). Neither the bridge nor Dasein (as an expression of its unique worldly situation) could appear without the fourfold always being together. The bridge is a bridge insofar as it delivers the peasant to a world of feudal commerce and medieval architecture, made possible by the tools that cut stone (but do not mix cement), the horses that move goods (but are not eaten), and the surplus that makes castles (but not factories). Any reconfiguration of these elements, either by the creative capacities of Dasein or the dynamic unfolding of the earth itself (for example, the emergence/discovery of new materials), unleashes potentialities that give rise to new things and new worlds. The fourfold is the totality of these shifting relations that variously allow the world to unfold as a world, that is, as a specific site marked out in and through its material emergence.

This brings us to the central concern of this chapter: How does the fourfold help us understand dwelling and, specifically, dwelling as a site of sovereignty? Indeed, I have just argued, mortals are no longer the agent of building but a participant. The fourfold situates Dasein "within the sway and dominion" of a world whose nature is to self-disclose. And while Dasein no doubt participates in the disclosive process, it is not its origin nor its final determinant. As W. McNeil (2006) suggests, dwelling is not "a human accomplishment, not something humans accomplish of their own accord or of their own merit" (p. 143). Rather, it is something we are delivered to and, in the midst of being delivered, we make decisions. That said, while dwelling in the fourfold is not an accomplishment or an achievement, it is still very much *a task*. By this I do not simply mean it is something that needs doing, but that its doing requires tenacity and sweat. It is a task undertaken in a world where Dasein's capacities are limited and its effects meagre. And it is a task that needs to be done again and again: "Dwelling is ... a challenge that has to be undergone ever anew" (W. McNeill, 2006, p. 145).

In understanding dwelling as a task that is never accomplished, we can begin to see the justification for characterizing dwelling as a claim. On the one hand, the world is full of projects that mortals build. Through the creation of poetry, maps, bridges and ontologies, mortals build the world they find. But as the discussion of *techne* illustrates, building is more than creating what we find. It means looking into the dark corners of being to build the world anew; to make the world mortals are in, their *own world*. In the words of Malpas (2006), "Building is the productive activity through which human beings make a place for themselves in the world and so by means of which their own dwelling is articulated" (p. 271). The key phrases here are "for themselves" and "their own dwelling." Building is about making the world we find our own world; that is, creating a place for ourselves through building. To be sure, the fourfold does not allow such activities any permanence. On the contrary, the structures that are disclosed through building are always already on their way to being overwhelmed by the fourfold's unfolding (by the trade routes that may circumvent the city, by the floods that may overwhelm the bridge, etc.). It is precisely for this reason that dwelling – as the practice of building one's world in the midst of the fourfold – can only ever be a *claim*.

On the one hand, it is a claim because, as Malpas (2006) suggests, building is how subjects endeavour to make a place for themselves. Thus, dwelling is a claim in the sense of planting a flag, as in "I claim this territory to be mine." This does not mean, to be clear, that such claims take shape through the language of possession common in the West and we should not be misled by our own ontological baggage in understanding this term. Indeed, Heidegger (2010) himself advocated for a modality of dwelling as "caring and preserving" where building was "power-free," leaving room for the earth to self-unfold (see Joronen, 2013; Ziarek, 1999) and there are many Indigenous ontologies that would claim the world in similar terms. Yet, I would argue, this is still a claim. To hold a place open in the fourfold that is power-free and preserving is claiming in the language of unclaiming, or what Heidegger (2010) calls "letting be." My point here is that even in "letting be," mortals build. They create a place where the world *can* be "let be" or, as Ziarek (1999) puts it, "Letting [be], while not entirely a human disposition or will *needs to be worked on* … it indicates that being transforms itself but cannot do so 'on its own,' without engagement, without human letting" (p. 182, my emphasis). Thus, even when humans are creating a space for care and preservation – a place where the earth can be *let* to emerge in and of its own accord – it is still something that needs to be built. It is still a claim, even if what that claim claims is to be power-free.

And yet, as already suggested, there is a second sense of claiming here: Claiming as an allegation. Every building (whether in the diminished form of modern technology or in the open form of letting be) emerges from and is delivered to the fourfold's unmasterable potential. It is the fourfold that establishes both the origin and limit of every claim. Even as every building emerges from the fourfold, it is that same potential that undermines its stability, and thus, maintains its status as a wager. The landscapes, totems, alters, and works of art that humans build do not simply express a world. Rather they express a world that unfolds, always moving and shifting under their feet. Even as they are built to solve practical problems, they also are built to address existential ones, namely, the problem of being in a world that is not mine, a world that precedes and exceeds me even as it is my origin. In this framing, building is about trying to establish some sovereignty over a world that perpetually eludes us. The structures we build are both practical structures and spiritual totems. Markers, placed in the earth, staking a claim.

Conclusions

The aim of this chapter has been to introduce a conception of dwelling as claiming. While the chapter inevitably covers some of the basics of Heidegger's thought, its ambition is to make two specific points. First, that dwelling describes how subjects create a home for themselves in the world and, second, how such homes are perpetually undermined by the world itself. In the discussion of *techne*, the home is understood primarily as an active, if not violent, struggle to create a place that is one's own; a place that builds the world one has inherited into something that is more than what one has been bequeathed. In the discussion of the fourfold, however, the emphasis is on the precarity of building. Even as the fourfold provisions mortal capacities, those capacities are tied to the fourfold's unfolding – a process that never ceases or stops. It is for this reason that the dwelling cannot be thought of as something built by a self-standing subject (i.e., a subject that stands on its own feet, independent of the fourfold's unfolding). On the contrary, the question of building is a question of how Dasein attempts to appropriate that which it receives. How mortals work to creatively enrol, and thus participate, in the world in which it is gathered. Taken together, I have argued that these two dimensions of dwelling allow us to see that dwelling is a claim. As a concept it stands for the ambition to mark out something that is *mine* in the midst of the unrelenting movement of the fourfold. It marks a desire to establish mastery over a world (even when the language of that mastery articulates it as unmastering). And yet, it also marks the futility of building in the fourfold. Claiming is always an ambition; a desire to stake a claim on a world whose parameters perpetually overwhelm. This is Dasein's destiny. The buildings that subjects build are only ever illusions of mastery – monuments to a sovereignty that never properly takes hold. Thus, to reassure ourselves with the height of our buildings, the speed of our motorways, or the sheer abundance of our material possessions, is to delude ourselves from reckoning with this most basic situation, that is, the situation of not being at home; of not having a place in the world that can resolutely be called *our own*.

Chapter Six

Dwelling as Both Open and Closed

Every dwelling is a fortress built to defend against the elements, a constant reminder of human vulnerability.

– Yi Fu Tuan, *Landscapes of Fear* (1979)

Introduction

The question I initially posed for this section is, "Why do subjects claim their world?" The answer, for Heidegger, is that subjects claim in order to make the world their own. Recall that Heidegger's subject is faced with an existential contradiction: the world is ours (in the sense that it defines who we are, establishing our parameters for being and thinking) and it is *not* ours (it is given to us by others). Dwelling is how mortals resolve this situation. It is by building that subjects endeavour to claim sovereignty over a world they are alienated from and overwhelmed by. Thus, even as subjects are always already in the world, that inheritance has no meaning until it is claimed. And yet, Heidegger also understands such claims as futile. They are assertions of sovereignty – not its consummation. While claiming is realized in material forms, those buildings are themselves always exposed to the fourfold, perpetually undermined by its unfolding.

As might be expected, I am not wholly opposed to this conception of subjectivity or the answer it provides to why subjects claim. I agree that subjectivity is essentially a struggle to take ownership over one's existence. I also like the sense of heroic futility in it, subjects trying to master the unmasterable, each our own Robinson Crusoe, building our small island kingdoms, destined to fail in our own way. But there is also something that niggles about it. The problem I have with Heidegger's (1971a) conception of dwelling is it conceptualizes this heroic struggle in a world always already conceived as *ours*. To be clear, I recognize that Heidegger understands true ownership to be achieved through building. But even with this being said, there is a more primordial conception of ownership subtending all and every endeavour to dwell. This is because, for Heidegger, being-in-the-world is always predicated upon a primary proprietary relation. One cannot be a subject without a world, a world that belongs to Dasein and that Dasein is obliged to build. But what if you are without a world? In Heidegger's view, the Jew is worldless (see Mendieta, 2017). Moving from place to place like hermit crabs, picking up the cast-off dwellings of others, Jews have nothing to care for or make meaningful through building. Thus, even as Heidegger is suspicious of worldly inheritance, and implores subjects to break through its everyday appearance, he nonetheless understands belonging as a prerequisite for any and every dwelling. And one should not take this notion lightly. For Heidegger (1962), there is nothing to be valued about building a home in the world of others. While he is no doubt disparaging of those that bury themselves in the tranquilizing power of the They (society, the public, etc.), he is even more disparaging of those that desire to flee their world, attempting to take on the history and language of others. Being-in-the-world is – in its essence – a problem of ownership. It is only by being-in one's own world that subjects can discover their ownmost being.

As already suggested in section 1, this project sets sail from a different premise. The world subjects are in (the habits we inhabit, the projects we embark upon, the tools that we take-up) is fundamentally not our own. We have no say over our difference. We have no possession over what we receive. Thus, even as Heidegger argues that we are thrown into a world that we had no hand in making, the condition of *Angst* emerges precisely because he perceives subjects as alienated from a world already presumed to be theirs. Dasein's pathology is driven by this presumption of primordial belonging. Sovereignty for Heidegger is not just an existential struggle, it is a moral one. It is Dasein's obligation to make its world its ownmost world. And this is because it belongs to Dasein and Dasein to it.

The question this chapter poses is whether it is possible to retain Heidegger's conception of subjectivity as a struggle for sovereignty while jettisoning the relation of primordial propriety. Can we think of the subject's heroic journey in terms other than the story of Odysseus and the Redcrosse Knight, returning to worlds that were always their birth-right? Perhaps we can think of it in terms of the wanderer – the Jew – seeking a place in a world? Even as the Jew has also been thrown into a world, delivered to a history, language, and tradition, this inheritance is not necessarily a source of anxiety and alienation. Indeed, it can also be a site of refuge and protection. Such a perspective opens up the possibility for considering a very different notion of dwelling; a conception I will elaborate through the work of Emmanuel Levinas.

Levinas is not particularly well known for his conception of dwelling (cf. Harrison, 2007b). Yet, I would argue that it plays a pivotal role in his conception of subjectivity. While this is most evident in Levinas's *Totality and Infinity* (1969), one could argue that it continues in his later work in the guise of different terminologies. The ultimate aim of this chapter is similar to the previous. It is to illustrate how subject's claim their sovereignty. But the way this claim is formulated is fundamentally different. By the end of this chapter, we will see that Levinas still understands the dwelling as something built and built to facilitate ownership and propriety. But we will also see that such buildings are illusory. The dwelling in Levinas is never properly a place. It is not even an event. While it names a site of ownership and self-possession, that site never properly occurs. In Levinas, subjects never inhabit the dwellings they build. This is not because they are interrupted (by the unfolding of the world) but because the buildings themselves are illusory, dreams of sovereignty permanently deferred. This is the critical difference between dwelling in Heidegger and dwelling in Levinas. With all the caveats of the previous chapter in place, Heidegger presents the dwelling as an event. It marks a claim to sovereignty that while precarious, temporary and threatened, nonetheless *takes place*. Subjects do dwell in Heidegger, even as its boundaries perpetually shake. Dwelling in Levinas never properly transpires. It is interminably unresolved, not just threatened but suspended. What the dwelling marks, therefore, is an ambition. It marks not a temporary claim but a desired one – a claim that (by its nature) is never realised. If Heidegger's dwelling represents claiming in one sense of the term (as in an assertion or requisition, for example, "he claimed the throne"), Levinas's dwelling represents claiming in another sense (as in a shaky proposition, for example, "he claimed to predict the future"). Dwelling in this chapter is still a claim. But it is formulated in a far more desperate and unconvincing key. While dwellings appear before us in their shining material light, their content lacks substance; their assertions elusive, if not dubious.

To illustrate this conception of dwelling, I have divided the chapter into three sections. The first section, introduces Levinas's conception of subjectivity, particularly as it takes shape in *Totality and Infinity* (1969). The central point here is that Levinas does not understand the subject first and foremost as "in-the-world." For Levinas, subjectivity arises out of a series of pre-ontological encounters with what he terms alterity. What he means by alterity, and how subjectivity takes shape through it, is precisely what will be explained. It is imperative to provide a brief sketch of these core terms for the remainder of the argument to make sense. The next section, moves on to explaining Levinas's notion of dwelling. For Levinas, the dwelling is an essentially paradoxical space.[1] On the one hand, the dwelling is the site that provisions subjectivity (it creates a site for the subject to find its own subjective capacities), but on the other, it is what brings the subject face-to-face with that which robs it of its sovereignty. As Levinas suggests, the dwelling is the home of subjectivity, a site where subjectivity is welcomed. But the home is wholly and interminably threatened, disrupted and haunted by alterity. The final section, develops this conception of dwelling further by introducing two terms from Levinas's later work: *the said* and *justice*. Taken together these terms extend our understanding of how the dwelling works to shield the subject from alterity. While the said does this by masking (by hiding alterity beneath the various ways it is thematized in language), justice does this through unburdening (transferring the weight of alterity to institutional arrangements). In either case, these two concepts emphasize the apparitional nature of the dwelling as well as its precarity in relation to alterity.

The chapter concludes by arguing that while dwelling is still the practice of claiming, what is claimed is significantly altered. For both Heidegger and Levinas, dwelling is always only a claim. But while Heidegger understands the precarity of the claim to be anchored in the unfolding of the fourfold, Levinas anchors it in a pre-ontological dimension that never lets the claim properly transpire. This means that the buildings subjects build (art, infrastructure, landscapes, etc.) are not simply unfolding (emerging and receding in the midst of the fourfold), they are illusory – that is, they never properly actualize the relations they claim. To understand this, however, we need to begin with Levinas's conception of the subject.

The Levinasian Subject

It is often said that Levinas presents us with a philosophy of ethics and this is no doubt true. But *Totality and Infinity* (1969) can also be read as a philosophy of the subject. And it is a philosophy that directly refutes the idea that subjectivity is founded in worldly experience. For Heidegger, what a subject *is*, is revealed through the possibilities of the world itself. Thus the subject grasps itself, and its potential, by engaging with the possibilities that the world reveals. Levinas, however, does not anchor the subject (or its possibilities) in the world. On the contrary, subjectivity is precipitated through its encounter with others; an experience that is both a concrete experience (it occurs through material engagements with other people) and an abstract one (it both subtends and transcends the specifics of any single encounter). Before the subject thinks, speaks or understands itself as a worldly being, that is, before any and all intuiting, naming or knowing, the subject is encountering others. This is how Levinas goes about changing the narrative arc of subjectivity. Rather than telling a story about how Dasein comes to see its situation and act upon its freedom, Levinas tells a story of the subject's *beholdeness*. While this story is also a story about the subject's efforts to be a subject, its narrative is definitively non-heroic. It is a story of incapacity and failure rather than of vision and action. But to see this we need to understand how the subject is formed.

There are two specific modes of alterity operating in *Totality and Infinity*: one encountered through the body and the other through the interhuman encounter (Large, 2015). In terms of the former, corporeality is Levinas's answer to Heidegger's conception of the care structure. In *Being and Time* (1962), Heidegger understands Dasein as a being that essentially cares for its being and does so by caring for the projects it finds in the world. For Levinas, however, caring is not a natural state. Subjects in Levinas do not come into a world of projects and labours they instinctively take up. On the contrary, the world first appears as a set of corporeal enjoyments. "We live from 'good soup,'" Levinas (1969) states, "air, light, spectacles, work, ideas, sleep, etc. ... we live from them" (p. 110). Enjoyment is the primordial experience of taking part in that which life offers without reference to a self. It precedes self-consciousness and self-definition. Rather than Dasein recognizing itself through its inherited world (and the self-conceptions it provides), the corporeal subject bathes in the world's qualities: the sun on its face, the drying wind, the coolness of water. The subject is first and foremost a body, transacting in life's elemental texture as pleasure and satisfaction. The body eats not only because it is hungry, but also because it enjoys eating. Rather than the world "being a system of tools, [it] is an ensemble of nourishments" (Levinas, 1987, p. 63). In enjoyment, the subject is corporeal and sensible rather than knowing and instrumental.

But enjoyment is also a double-edged sword. The problem, as the body soon discovers, is that the enjoyments life provides are not under its dominion or sway.[2] It is here that something like Heidegger's care structure enters the fray, though in a very different way. *Caring* for Levinas (though he does not use this term) is instigated by a recognition of corporeal insecurity: "nourishment comes as a happy chance," Levinas

(1969) states, "on the one hand [it] offers itself and contents, but [on the other, it] already withdraws, losing itself in the nowhere" (p. 141).[3] Even as the body enjoys the cool breeze, it senses that the source of that enjoyment is irretrievably outside its power. That which the body enjoys "come[s] from nowhere," Levinas states, "appearing without there being anything that appears – and consequently coming always, without my being able to possess the source" (p. 141). This event forces the body to reckon with the precarity of its enjoyment. The body is uncertain about its pleasures: "The uncertainties of the future ... remind enjoyment that its independence envelops a dependence" (p. 144).

The significance of this reckoning is it opens up within the subject a dimension of interiority.[4] This dimension can be compared to the one precipitated by Heidegger's broken hammer. In Heidegger's (1962) parable, Dasein is using his tools in an absorbed manner not thinking about what the hammer is or how it works. But when the hammer breaks, Dasein needs to consider it as an object for contemplation and concern. Similarly, Levinas's (1969) body enjoys the elemental qualities of the world without thought of their continuation. But when those elements disappear, the body concerns itself with their absence. The key difference is that while Dasein can repair the hammer, the subject cannot summon the wind. The wind is beyond her. It exists in a dimension whose origin and source is outside her, beyond any capacity she could wilfully summon.[5] The subject thus emerges not as a being that can manage and control its world (by fixing the hammer) but a being that recognizes its exposure and beholdeness to forces outside it. "Enjoyment reveals the body's reliance on an absence," Levinas (1969, p. 142) states, and it is this revelation that precipitates a certain kind of ego to take shape; an ego sensitive to, and cognizant of, its exposure to alterity.

Thus, we can see how Levinas situates the self-conscious subject in very different terms than we normally see in the Western tradition. Most obviously, Levinas does not position self-consciousness as something always already there, intrinsic to our being and (in Heidegger at least) attuned to a world which it intuitively knows. On the contrary, the self-conscious subject is *solicited.* The fact that the wind disappears *forces* the ego to emerge.[6] While there is a proto-subject that precedes this interruption, it is defined in corporeal and sensible terms rather than through knowing and understanding. It is only when enjoyment is lost or delayed that consciousness emerges. To be clear, there is nothing uniquely human about this conception of subjectivity. Indeed, consciousness, in this mode, is conceptualized as a capacity to be "concerned for the morrow." And while I am not an expert on animal consciousness, there seems to be plenty of prima facie evidence that animals summon all kinds of (bio-chemical, bio-mechanical, neuro-cognitive) resources to address similar concerns (e.g., building nests, storing food, using tools). It is at this point that Levinas introduces another encounter, an encounter that is uniquely human and pushes the ego across a different threshold of consciousness and self-hood. Understanding this encounter and how it facilitates the subject means engaging with what is perhaps the most well-known – and most important – dimension of Levinas's work: the arrival of the other human being.

Before the arrival of the other person there is no reason for the "I"-ness of the subject to be thought. There is simply the body enjoying what it finds. This body does not need to identify itself or name itself. It may need to plan for its future enjoyment, but it does not require a conception of itself as a singular self-conscious being. On the contrary, it only needs a certain conception of futurity and survival. The arrival of another person, however, does solicit a different modality of self-hood; it forces the ego to think of itself in new and different terms.[7] For Levinas (1998), this insight stems from the basic empirical facticity that another person demands a response: "In every attitude toward the human being there is a greeting – even if it's the refusal of greeting" (p. 7).[8] On the one hand, the other person instigates a range of social relations. Whether she is a work colleague, a friend, a police officer, or an elected representative, she solicits a network of understandings, expectations, and connections that help me know who she is and the role she plays in my life. But on the other hand, she also exceeds these relations. She is more than a set of roles. As Levinas suggests, there is always a dimension of the other human that resists such classifications; that challenges "all the definitions or descriptions the world gives" (Cavell, 1999, p. 390). Thus, "the relationship with the other is not an idyllic and harmonious relationship of communion, or a sympathy through which we put ourselves in the other's place" (Levinas, 1987, p. 75). On the contrary, the other's resistance to being subsumed by the social relations in which she appears opens up another kind of relation, one where I recognize the other as a mystery: "We recognise the other as resembling us, but exterior to us; the relationship with the other is a relationship with a Mystery" (Levinas, 1987, p. 75).

In many ways Levinas's (1987) philosophical project can be characterized as an attempt to illuminate the many ways that "Mystery" interrupts being. He argues that death is mysterious not because no one has experienced it but because it *cannot be experienced*. It is, "*ungraspable* ... it marks the end of the subject's virility and heroism" (pp. 71–2, my emphasis). He argues that time is mysterious as it denies our understandings and expectations in regards to the present. As already discussed in chapter 4, the future is not something humans can grab hold of, on the contrary, "the future … lays hold of us" (p. 77). Most importantly, Levinas understands other people as mysterious. While we endeavour to understand each other, govern each other, love each other, and change each other, other people perpetually surprise us, upending predictions and defying expectations. It is this dimension that makes our relations with others both immanent and transcendent (Large, 2015). In terms of the former, we address others through language and representation, "the cultural meaning which is revealed … as it were horizontally, which is revealed from the historical world to which it belongs and which … reveals the horizons of this world" (Levinas, 1996a, p. 53). But "this mundane meaning is disturbed and jostled by another presence that is abstract … and not integrated into the world" (Levinas 1996a, p. 53). This is the transcendence of the other, the aspect of the other that arrives from the "dimensions of height" (Levinas, 1969, p. 155).

It is this dimension (beyond the mundane horizons of the world) that keeps each and every human being unique and irreplaceable and what makes every attempt to colonize, enslave, or control the *soul* of another (i.e., the other person in their entirety) impossible.[9] Most importantly for our purposes, it is also this dimension that allows the "I" to surface in its distinctively human form. If the other could be subsumed by me – if it could be wholly subjugated into my sameness, brought fully under my control, integrated wholly and completely into my economy of enjoyment – then there would be no need for an individual or you to arise. In other words, self-consciousness is instigated by the recognition of a gap, a separation between us that is non-traversable. It is precisely because the other *resists* me – they resist my enjoyment, my understanding, my control, and my love that I come to see myself as an "I" and you as a "you." This is why the other's alterity matters. It is because the other is *other* that I am summoned. It is because the other exists at a distance that I am prompted to ask, "Who are you?" It is this unthematizability – this resistance – of the other that gives my ego cause. As Harrison (2007a, 2007b) suggests, the other opens an unbridgeable space between us; what he terms "a relation of non-relation," whereby the "I" is dependent on the other's arrival and (simultaneously) the other's infinite distance, in order to be a subject.

It is only at this point that we can begin to see why Levinas characterizes subjectivity as *beholden*. I would not be capable of thinking myself as a self if the other had not arrived, interrupting my enjoyment and precipitating my consciousness. In "putting my consciousness in question," Levinas states (1996b), the other gives me the "consciousness of putting in question" (p. 16). The other, therefore, is the origin of my "I'ness." I would not be me if she did not arrive. Thus I am beholden through and through not only to her existence but to her distance, that is, to her alterity, that which resists me and (in resisting) solicits my attention.[10] It is both her *and her distance* that is the foundation of my being. In addition, we can now see why Levinas characterizes this relationship as fundamentally *ethical*. Because my uniqueness as a human depends upon the uniqueness of the other, I am obliged to take care of her (and her uniqueness) from the start. My individual being begins with my responsibility to you: "A responsibility that, from the start, devolves upon the *I* in the very perception of the other ... a responsibility that cannot be refused" (Levinas, 1998, p. 227). This characterization of responsibility, as one that cannot be refused, is key. I cannot choose responsibility. On the contrary, in the very event of becoming me, I am chosen. Subjectivity, for Levinas "is being *hostage*" (Levinas, 1981, p. 127). Ethics is rooted in the recognition that the dimension of myself that I take to be most my own (my life, my self, my consciousness) belongs to the other, a being over which I have no command. As a relation of non-relation, the other's power over me is total and non-equivalent: "In the very heart of the relationship with the other that characterizes our social life," Levinas (1987) states, "alterity appears as a nonreciprocal relationship" (p. 83). Thus, even as the other is the origin of my subjectivity, I have no claim over her. This is the double bind of subjectivity. My existence is fundamentally reliant on that over which I have no say. And it is this situation that leads Levinas to reconfigure Heidegger's concept of the dwelling.

Levinas's Dwelling

Before delving into the profound differences between Levinas's and Heidegger's conception of dwelling, it is useful to consider where there is agreement. First of all, Levinas, like Heidegger, understands the dwelling as a site of self-possession, it "expresses my place on this earth ... it is the very boundaries of my subjectivity and without it I am hardly a self at all" (Large, 2015, p. 55). Second, the dwelling is understood as something built. As I discuss below, the dwelling is something actively cultivated and cared for in the face of the other. Finally, and most importantly, the dwelling in Levinas is also thought of as a home, a place where the subject endeavours to establish a place for itself in the world. The key difference for these thinkers is how they conceptualize the forces of alienation and alterity against which the dwelling is built. For Heidegger, those forces reside in the world itself, that is, in the alienating nature of inheritance (in his early work) and the overwhelming unfolding of the fourfold (in his later work). For Levinas, however, alterity is always the other. It comes not from the world itself but from outside it. While this distinction seems minor, it fundamentally changes how we understand what it means to dwell. Both authors use the term to describe how subjects build themselves a place in the world. But for Heidegger, dwelling is an event (even if it is a short lived one) while for Levinas it is a desire (one that never completes). To understand this distinction, we need to understand a key paradoxical phrase that Levinas (1969) uses to describe the dwelling. The dwelling "must … be at the same time open and closed" (p. 148).[11] The remainder of this discussion endeavours to explain and extend this formulation.

When Levinas (1969) states that the dwelling is closed, he means that it operates as a site of interiority, meaning a site that establishes the boundaries of "me" versus "you." As already suggested, the inter-human encounter precipitates the advent of self-consciousness, that is, the recognition that I am more than a sensible body seeking to secure its enjoyment. I am also an "I" seeking to define and secure myself vis-à-vis other human beings. There are three characteristics Levinas develops to explain this site of interiority. First, Levinas describes it as a place where the subject recollects itself. On the one hand, he means this in the normal way. It is where the subject remembers herself as if waking-up from a reverie or a dream. But he also means it in the sense of being collected, as in pulling one's self together, assembling oneself into something cohesive and unified, an event of self-possession. In recollection, the ego discovers not only that it is there, but that its thereness endows it with certain powers and capacities: it "designates a suspension of the immediate reactions the world solicits [in enjoyment] in view of a greater attention to oneself, one's possibilities and the situation" (Levinas, 1969, p. 154). This means the ego discovers that it can do things: "dwelling is the very mode of *maintaining oneself*," it is discovering that it is a "body that … holds *itself* up and *can*" (Levinas, 1969, p. 37, original emphasis). Within the dwelling everything looks open to the subject, available and awaiting its attention, knowledge, and power: "Everything is at my disposal, even the stars, if I but reckon them … everything is here, everything belongs to me" (Levinas, 1969, p. 37).

This leads to the second characteristic of its interiority. The dwelling is a site of representation.[12] To be sure, we have to be careful with this term. The point is not that the dwelling is a representation itself but it is where representation happens. Here, Levinas (1969) is drawing attention to the inherent sociality of the dwelling. As he states, in "the interiority of recollection," the subject discovers "a world already human"

(p. 155). It is in the dwelling that the subject discovers its history, its language, its culture, and all the representations that bear the traces of those that came before me. The metaphor Levinas returns to here is the home. The dwelling is where we find the table set and the bed made. It is a place of rest, respite, and comfort where representations operate as a resource for me to use in my ambition to speak and describe myself as a subject through the terminology I have inherited.[13] Such a description hews very close to Heidegger's concept of the world. Yet, rather than the world being immanent to the subject's existence (the ground of its being), Levinas understands it representationally. This is a position Heidegger would obviously reject. For him, the world is ontological and the subject does not "find" it but is thrown into it. The journey of Heidegger's subject is learning to see this world as something it can tear from history and make one's own. Levinas, however, does not see the world as alienation but as welcoming. It is what provisions the subject, granting a place of refuge where it can partake of the comforts and securities provided by those that came before. Meaning systems, tools, established social roles, institutions of law and order, all these inherited forms are not seen as corrupting but as welcome resources for defining and understanding who I am. They provide a means for the subject to secure itself through well-worn comforts.

This leads to the final characteristic of the dwelling's interiority, which is that it is a site of labour and work or what Levinas (1969) calls *economy*. As already suggested, in the dwelling, the ego discovers a certain power over the world. In doing so, the enjoyments the body once sensibly experienced now appear as needs it can work to secure: "to be cold, hungry, thirsty, naked to seek shelter – all these dependences with regard to the world, having become needs" (Levinas, 1969, p. 116).[14] As needs, the dwelling comes to be a site of calculation and measurement. And the ego pursues these needs with the intent of mastery: "By my labour I must secure a place for myself, one to which I can retreat from the wind and the sun ... into that space I take what I require from my world, thus transforming beings into possessions. This taking possession opens a masterable time in the future, the time in which I may enjoy or replenish my stocks" (Bergo, 2003, p. 69). The subject has thus travelled far from being a body in enjoyment. In the dwelling, the subject discovers clock time and works in the now to save for tomorrow. It is also where resources are gathered and saved.[15] Thus, while the home is a site of welcome, it is also a site of industry. It is the place where the subject manages its life in a systematic and calculated manner: "With labour, the ego fixes things all around it, structures them, makes them property. This mastery of the elemental blocks its original substancelessness, its unforeseeable futurity, its abyssal hold over me" (Blumenfeld, 2014, p. 114). Thus, it is through labour that the dwelling's interiority is both found and built.

In sum, the dwelling is a site of self-mastery: a place where the subject finds the resources it inherits and uses them to establish its freedom and secure its comfort. As a site of interiority, everything in the dwelling appears to be under its command. Yet, it is precisely at this point that Levinas emphasizes the illusory nature of the dwelling. Let us use my own typical middle-class life in the UK as an example. As a white privileged citizen, I am situated within a network of institutions, knowledges, and norms that collectively work to keep my life secure. Whether these be economic systems (grocery stores and housing markets), health systems (hospitals, medical research, and public health bodies), legal systems (the police, the courts), or cultural systems (norms of civil conduct and social justice), I am embedded within representational networks that reinforce the idea that I live a sovereign life – a life that appears as if it were my own. If I get sick, I go to the doctor. If I am hungry, I go to the store. Everything is there, available for me to live the life that is mine. And yet, Levinas reminds us that none of these systems make alterity go away. The fact that I am surrounded by available food does not mean I don't get hungry. The fact that I can visit the doctor, does not mean I don't get sick. Our food systems and health systems are masterfully organized economies of human ingenuity. But they are designed to *respond* to the existential problem that living poses. The problems themselves remain – addressed but not resolved. Indeed, the fact that they are addressed in a manner that is uneven, unjust, and keep some bodies more vulnerable than others, is precisely because all beings begin from a position of vulnerability. As Large (2015) suggests, the very fact that humans have bodies means that their lives are dependent (utterly and wholly) on situations outside them: "The existence of my body … demonstrates that I am dependent on the exteriority of the world" (p. 57). This is the conundrum of being a living being. We are exposed to problems that we cannot bring under our dominion or control. We are vulnerable, through and through, to that which is outside us.

It is for this reason that Levinas (1969) characterizes the dwelling as a site that is both open as well as closed. While it provides the necessary interiority for the subject to define itself as a sovereign and separate being, that provision is predicated upon a primordial exposure to alterity. Whether that alterity is conceptualized corporeally (our body's interminable exposure to hunger, disease, and the mysteries of death and the future) or cognitively (through the face of the other who precipitate's our capacity to name ourselves as an "I"), the dwelling's emergence is anchored in these mysteries. This is why the dwelling cannot be conceived *only* as an interiority. To do so would be to suggest that the subject is actually sovereign - that it achieves the freedom and mastery it seeks. But the whole point of Levinas's metaphysics is that this can *never happen*. Subjectivity is, by definition, wholly and permanently *beholden*. We have no choice about this. There would be no dwelling if the other did not give it cause. This is to say that the other must still be there, haunting the dwelling, permanently revealing that its interiority is an illusion; a representational economy that masks alterity but by no means eradicates it. It is for this reason that Levinas's dwelling cannot be considered an event. The dwelling never properly transpires. To suggest that it is *there*, even momentarily, would be to suggest that the dwelling is closed – that the interiority it represents is or can be achieved. But the dwelling is always both open and closed, its closure permanently deferred. Better to think of its interiority as representational; a fabrication needed to exist in the face of alterity; an economy that has no force, no capacity, to secure the subject in the interiority it claims.[16]

In sum, to sit with Levinas's (1969) thought is to sit with the paradoxical space of the dwelling. To characterize the dwelling as a site of freedom, mastery, identity, and economy is no doubt true. The dwelling provides a site where "everything is here, everything belongs to me" (p. 37). Whether we are talking about the advances of modern science or the healing power of the shaman, the dwelling is where human ingenuity (in all its creative diversity) brings the world to light. But this mastery is also illusory. The dwelling is built in response to the one who gives freedom and "like a thief in the night" steals it away. Regardless of how many hospitals, satellites, totems, and food systems we build, we are still beholden. Just behind the façade of being, the other awaits, interrupting every sense of mastery and self-possession. Does this mean that the dwelling is simply a façade? No. For Levinas, the dwelling is the site of all sociality, the place where real social relations happen. But what makes the dwelling work is the fact that it keeps alterity *hidden*. It is this dimension – this dimension of hiding and forgetting – that needs to be developed if we are to understand the dwelling as a claim.

The Mask of Language and Justice

In the previous section I referred to the interiority of the dwelling as illusory. My point was to draw attention to the fact that the representational economies that humans see and operate within are fabrications – things built to hide the haunting, unpowering problems that alterity poses. The aim of this section is to think through this illusory representational economy more carefully, and in doing so, establish the underlying theoretical position informing this book: the proposition that culture is never properly present, but rather, is a dream of presence; a representational economy necessitated by a prior existential demand. Specifically, this means spelling out more cautiously what I mean when I use terms like *illusory*, *hiding*, and *masking*. The problem here is that there is a long history of social and political theory dedicated to unmasking. From Marx's notion of demystification to Foucault's unmasking of the relation between truth and power, the social theorist has a long and celebrated tradition of exposing the unrecognized social relations within which we live. The point I am making here, however, is that appearances matter. Rather than approaching appearances as fake, false knowledge or duplicitous truth, appearances constitute the social relations in which we must (with the emphasis on *must*) live. As Levinas (1998) suggests, appearances structure the sphere of intelligibility: "The sphere of ... in which everyday life as well as the tradition of our philosophic and scientific thought maintains itself, is characterised by vision [and] the structure of seeing" (p. 159). This is why the representational economy is characterized in terms of seeing and light. It works by hiding alterity's debilitating darkness, shrouding it in visibility. It is "the privileged status of representation [that] makes us forget," Chanter (2001, p. 64) states, and it is only by forgetting alterity, by not seeing its incapacitating darkness, that the subject can imagine itself – can dream itself – as sovereign. In order to illustrate this dimension of the dwelling (this capacity to mask and hide through light), this section examines two further concepts – the said and justice – both of which emerge in Levinas's *Otherwise than Being, or Beyond Essence* (1981). Taken together, these concepts demonstrate how dwelling works by concealing alterity, that is by shrouding what Levinas calls "the hither side of ontology": that dimension beyond being whose shadow is covered over by the light of representation.

The first concept (the said) emerges from Levinas's critique of Heidegger's approach to language. For Heidegger, language is the means through which the world reveals itself (J. Powell, 2013b). It is also one of the means by which mortals break through everyday appearances.[17] Nowhere is this more clearly stated than in Heidegger's essay "... Poetically Man Dwells ..." (1971d). On the one hand, he argues that language allows mortals to illuminate the familiar parameters of their being: "What is familiar to man [is] everything that shimmers and blooms in the sky and thus under the sky and thus on earth" (p. 225).

The poet, however, uses language to reveal what is hidden and dormant in the world's appearance: "The poet, if he is a poet, does not describe the mere appearance of sky and earth. The poet ... causes the appearance of that which conceals itself" (p. 225). As discussed in the previous chapter, the poet uses language creatively, exploring its dark corners to reveal something new or previously unseen within the world. To dwell poetically, therefore, is to stretch the ontological breadth of being. Even as "everyday talk" masks the world and its potential, the poet can break through the shadows, bringing "the brightness and sound of the heavenly appearances into one with the darkness and silence" (p. 226).

In some ways Levinas's notion of the said starts from a similar place.[18] For Levinas, language belongs to the realm of the visible in that it operates through vocabularies that the subject finds already resident within the dwelling. Indeed, the past tense of the said is key. The said is that which has been already said. It is a world we find with speech already in it, sayings, phrases, modes of articulation, which give the subject the potential to speak: "It is through the already said that words, elements of a historically constituted vocabulary, will come to function as signs and acquire a usage, and bring about the proliferation of all the possibilities of vocabulary" (Levinas, 1981, p. 37).[19] Given this rendering, it is not surprising that Levinas appropriates Heidegger's notion of ontology to describe the said. Like ontology, the world, the clearing and other co-extensive Heideggerian terms, the said similarly describes a site of seeing and knowing. The difference, of course, is that Levinas locates the origin of this ontology in a more primordial domain. As Chanter (2001) puts it, Levinas "seeks not to deny the validity of ontology, but to go beyond it" (p. 147). This "beyond" the said is what Levinas calls *the saying*. The saying is everything that cannot be said. The emotional pain that cannot spoken. The searching for words as they stall and stutter. These are sayings that cannot be said. And yet they provide the gravitational force for every attempt at speech. It is for this reason that Levinas (1981) argues that language is not *nominational*, that is, it is not an attribution of words to ideas, phenomenon, and sensibilities already present in our visible world. Rather, it is *propositional*: "Language can be conceived as the verb in a predicative proposition" (p. 40). Every articulation of the said is an attempt – something proffered – trying to give words to that which escapes language. It is for this reason that Levinas refers to every said as "abusive," "monstrous," "profane," and a "betrayal." And yet such betrayals are necessary: "Being and manifestation go together," he states, "it is in fact natural that if the saying on the hither side of the said can show itself, it be said already in terms of being. But is it necessary ...? *It is necessary*" (p. 43, my emphasis).[20]

There are two key points here I want to draw out. First, while the said is being thought of as that which betrays, abuses, and profanes saying, it can also be thought as that which *hides* it. It is the said that we see and that makes the world visible. For Levinas (1981), all that we normally take to be language – from screaming rants, to soaring prose, to scientific exposition – is a collection of failed and scattered remnants of the saying. What we do not see – what remains hidden on the hither side of the said – is what speech fails to make visible. The failure of speech is what *does not show*. What shows is more and more speech; more and more renderings of the saying in ever new forms of exposition. This is the said perpetually endeavouring to capture all that is lost, missed, or inadequately described. Thus, all that stands before us – all that is visible and present; all that we correlate, debate, and categorize; all that we take to be most ontologically *real* – hides the primordiality of the saying.[21] The second point, is that this betrayal is *necessary*. This is a point I will return to throughout the book. Without this betrayal of the saying there would be no speech. Indeed, there would be no way of accessing the saying at all. The only way we can hear the unsaid is through its betrayal – through the practice of striving to speak the unsayable, even as it withdraws (interminably) from language. Such is the nature of the sphere of intelligibility in which we live; a sphere where betrayal is necessary (is it necessary? It is necessary). This leads to the second term I want to discuss – Levinas's conception of justice.

While the saying and the said have been a topic of some debate within scholarship on Levinas (Bergo, 2003; Chanter, 2001; Critchley, 1999a; Hutchens, 2004; Large, 2015), Levinas's writings on justice have had a much wider currency, specifically in discussions of ethics (Critchley, 1999b; Ruti, 2015) and political theory (J. Butler, 2004, 2005; Derrida, 1992, 2001; Žižek, 2005). Often these debates surround the ambiguous dynamic between the institutions of justice – courts of law, law schools, traditions of legal analysis, and so on – and the ethical responsibility that informs and transcends them. While these debates are interesting, my interest is in how these institutions work to representationally unburden the subject from its ethical responsibility, that is, cover over the subject's ethical exposure. To see this, however, we need to understand a specific problem Levinas addresses in his discussion of ethics. Recall that for Levinas, I am infinitely responsible to the other. As the instigator of my self-definition (my "I"-ness), I am obliged to those others (neighbours, friends, and strangers) who call me forth. But what about *their* neighbours, friends, and strangers? As Levinas (1981) suggests, to be responsible not only for my neighbour but for their neighbour and their neighbours ad infinitum "put[s] distance between me and the other" (p. 157). How can I be responsible to the one and simultaneously be responsible to the all? For Levinas there is no distinction. I am equally obliged to the human in front of me and those suffering on TV. But how is this possible?

Levinas's concept of justice illuminates a very specific type of economy that resides in the dwelling: an economy that weighs, calculates, and decides how to *distribute* responsibility.[22] Thus, the question of justice introduces "an incessant correction of the asymmetry of proximity" (Levinas 1981, p. 158). This correction is a move from inequality, where I am beholden to the other without reserve, to equality, where "I am approached as an other by the others" (p. 158). For Levinas, this transformation marks the birth of a democratic, liberal, and judicious economy, an economy based upon relations of reciprocity and equality rather than inequality and beholdeness. In this representational system, "everything is together, one can go from the one to the other and from the other to the one, put into relationship, judge, know, ask 'what about ...?'" (p. 158). The key point here is that this economy (call it justice, law, good government, etc.) *suspends* infinite responsibility so that judgments can be made. As Peperzak (1993) suggests, it is the appearance of many others that forces the subject to "temper ... the hyperbolic intensity of transcendence and bring it under the measure of a well-balanced justice" (p. 183).

To be clear, Levinas (1981) continually reinforces that infinite responsibility is primary: "In no way is justice a ... degeneration of the for-the-other, a diminution, a limitation of anarchic responsibility ... the equality of the all is borne by my inequality, the surplus of my duties over my rights" (p. 159). Thus, it is the primordial condition of beholdeness that allows an economy of equality and justice to take shape: "My relationship with the other as neighbour gives meaning to my relations with all the others. All human relations as human proceed from [this] ... justice is impossible without the one that renders it" (p. 159). And yet, in order for that ethical obligation to be operationalised, it must be reduced (like the said) into a representational economy. A system that not only calculates and measures the incalculable (remember that ethical responsibility is infinite), but hides its debilitating power. Ethics, in other words, is hidden on the hither side of justice.

It is in this sense that systems of justice can be thought of as an unburdening of the subject. In its capacity to represent the subject as equal and take over its ethical obligations, the representational economy of justice makes ethics (in a compromised manner) possible. "The multiplicity of human individuals causes a perpetual conflict," Peperzak (1993) writes, "unless an effective administration of justice prevents or conquers a general way by imposing many compromises" (p. 184). Thus, even as

Levinas (1981) insists (repeatedly) that these compromises are informed by a primordial obligation, that obligation is easily forgotten behind the institutions administering ethical responsibility on our behalf.

Once again, we see this tension between what is transcendent and real (the ethical obligation which is unpowering and dark) and what is necessary – the thematization of that obligation into economies of justice. Even though this thematization "distances me from the infinity of my responsibility" (Peperzak, 1993, p. 181), and subsumes "particular cases under a general rule" (Levinas, 1981, p. 159), Levinas (1981) reiterates that "justice *requires* contemporaneousness of representation. It is thus that the neighbour becomes visible" (p. 159). In other words, there *must be* a said of ethical responsibility; a said that not only economizes the work of ethics but does so in a manner that *discharges* the ethical obligation on our behalf, that is, that effectively unburdens subjects from ethical responsibility. As Peperzak (1993) puts it, if ethical responsibility is to mean anything at all, it must mean this betrayal: "The irruption of the ethical into the anonymous is not possible unless those who are facing and speaking to one another also participate in the interplay of the political and economical" (p. 179). It is necessary for the ethical obligation to politicize itself and in doing so represent itself as justice. Thus, even as Levinas (1981) repeatedly reminds us that this representation merely hides, rather than supplants, the ethical obligation, the condition of dwelling, as Chanter (2001) suggests above, *is forgetting*. Regardless of how often we recognize injustice disrupting justice, the solution is always more justice: more laws, better governance, more equitable calculations and distributions, more efforts to resew the tears that ethical inequality creates. Even as we may hear the call of the other, it is justice that appears and makes the ethical obligation manifest. For Levinas (1981), this is the very definition of society, or as Peperzak (1993) expresses it: society is a "collectivity whose social structures *hide* the unique and personal character of the face-to-face relation among its members" (p. 179, emphasis in original). Is this hiding and forgetting necessary? It is necessary. Alterity can only be a subtractive power. It is not a force, a capacity, a will, or a relation. Thus, being a subject means forgetting; it means entering into a representational economy that allows subjects to be subjects.

Conclusions

It is at this point that we need to return to the question of dwelling and specifically to the idea that dwelling is a claim. As should be clear by now, Levinas provides a different conception of dwelling to that proposed by Heidegger. And while dwelling is still being thought here as a claim, it is a conception of claiming that looks quite different. But to see this, let's briefly review the section as a whole.

In the previous chapter, I argued that Heidegger's conception of dwelling could be conceived through two key ideas: *techne* and the fourfold. On the one hand, dwelling is an event of building. It is a domain that mortals build in order to express their potentiality vis-à-vis the world into which they are thrown. But on the other hand, buildings are essentially precarious, always on the cusp of being overwhelmed by the forces of the world itself. Thus, while the dwelling can be thought of as a site of sovereignty (or at least a kind of sovereignty), it is not one with any resilience. It is for this reason that I characterized dwelling as a claim. While the buildings, monuments, and landscapes that mortals build appear permanent and powerful, they should not be thought of as expressions of actual social relations. They are claims. It is true that the bridge materially connects the world in innovative ways. But the relations around it rise and fall. Its claim on the world is temporary and destined to be undone. Buildings never achieve the sovereignty they stake. Like the king claiming his colony by virtue of a map, buildings have no proper purchase on the world they claim.[23]

The concept of dwelling in Levinas is very different. For Levinas, subjects are not primordially conceived as in-the-world. They do not get lost in conditions of alienation nor assert themselves through gestures of ownership. On the contrary, subjects emerge through their encounters with others. Other conditions, other times, other events, and other human beings urge subjects into existence. Such a conception of the subject fundamentally reworks our initial conception of dwelling. The pivotal idea for Levinas is that the dwelling is simultaneously open and closed. It is closed because it situates the representational economies needed to establish a predictable and unburdened life. And yet, the security it provides is illusory – an appearance that is pre-undermined by alterity. While the dwelling makes subjectivity possible, the world it circumscribes does not extend beyond the representational edifices it claims. The dwelling is always open and closed. The former remorselessly taking away from the latter.

So how does this conception of dwelling reformulate our conception of claiming? In the previous chapter, dwelling was conceptualized as the means by which subjects claim their world as their own world, that is, a world that they (creatively, poetically) wrest from their inheritance and the fourfold in order to create something that is theirs. But it is precisely this inclination towards ownership and mineness that Levinas is attempting to undo. For Levinas, the dwelling is never ours. While it appears as ours, due to the representational economies it fosters, this ostensible sovereignty is permanently exposed to alterity. It is for this reason that dwelling is, and always must be, a claim. But this is a different kind of claim than I discussed in the previous chapter. For Heidegger, claiming is an event – it transpires in a platial sense. For Levinas, however, dwelling is not an event. This is because dwelling is asymptotic, infinitely deferred. It can never properly close. Thus, while dwelling is claiming in the sense that it marks the subject's ambition to be a sovereign subject – to have a language, history, and ontology that is *its own* – it never properly *takes place*.

In conclusion, both Heidegger and Levinas present the dwelling as something built. The claims that emerge from the dwelling appear in clay and light, shining before us in their various material forms. But for Levinas, these appearances do not represent actual stakes in the ground. On the contrary, they are only ever claims. They mark the subject's desire to be a subject in a world where subjectivity is perpetually stolen. While the dwelling situates a site where sovereignty appears, we should not be seduced by those appearances. As already suggested, sovereignty for Levinas is not an event, it cannot transpire. But it can appear to transpire. As Chanter (2001) suggests, the dwelling is a place where the subject forgets. It appears in surfaces and light, obscuring the dark alterity which is its cause. And while this "hiding through light" allows subjectivity (as sovereignty and freedom) to appear, it *does not make alterity go away*. On the contrary, alterity always bursts through: in the future we did not see coming, in other people who refuse to be influenced or controlled, in the calamity that shakes all our representational edifices – all our reliable systems – to the ground. The dwelling is never closed. Thus, rather than conceptualizing dwelling as a poetic construction, I have rendered it here as a necessary illusion – that is, a claim. It is the claim that subjects require to exist. Being in the world (no hyphens) means making claims.

To wrap up this section, we can hopefully begin to see what I mean when I conceptualize culture as a claim. Culture names the representational edifice that subjects claim in order to dwell, that is, in order to be subjects in the world. While section 1 sought to separate the question of difference (which is unknowable), from the question of claiming (why people claim the differences they have), this section has theorized the phenomenon of claiming itself. In doing so, I have established a path for conceptualizing culture not as a set of present differences but as a set of representational claims. Culture, in this framing, constitutes the illusory representational economies subjects build to dwell in the world. It is not a modality of difference that is present, but a modality of difference that is dreamed as present. Understanding what this means and how claiming can be applied to a theory of culture and cultural geography is the topic of the next and final section.

SECTION THREE

Geography

Figure 3. *Spring (The Earthly Paradise)* by Nicolas Poussin, c. 1600–1664.

Dreams of Ownership

In Nicolas Poussin's painting *Spring (The Earlty Paradise)*, we see two figures – Adam and Eve – in the centre of an opulent and abundant landscape. While the figures are central, situated in the focus loci of the painting, they are not dominant. Rather than standing out and apart from the wilderness, they easily blend in, not so much masters of this new world but part of it, caught up – if not subdued – by its density of vegetation and shadow. These are not the cut porcelain figures of Durer, Michelangelo, and Titian. On the contrary, they are wild and contorted creatures, still wet from the earth and clay from which they were made. But this is all about to change. Eve is pointing Adam to the forbidden tree and God is in the top right-hand corner in retreat. The painting's title is associated with the pagan goddess Flora, as if these not-yet-quite humans are about to emerge from their chrysalis. While the coming exile is tragic, it is also what separates them from this dense domain of earth and root, this world they ostensibly rule.[1]

While we may not often think of Eden as an island, I present it here as another example of the sovereignty ideal, a site of total ownership and dominion by two humans cut from the same cloth – a pair that is also one. As Derrida (2009) points out, this is Rousseau's dream of democracy. A system where the people and the state did not need to bother with the tedious work of negotiating their differences because – for all intents and purposes – they are the same: "a country or a state in which sovereign and people would be a single person" (p. 22), one reflected in the other, "the insular utopia of an individual alone enough on an island to be both the sovereign and all the people gathered together" (p. 22). Is such a dream not Edenic? In giving Adam and Eve dominion over "the fish of the sea, and over the fowl of the air, and over the cattle, and *over all the earth*" (Genesis 1:26), their mastery is characterized not as an inheritance over something already there (with all its content and legacies waiting to be navigated and negotiated) but as something created for them, an inheritance shaped in order to be bestowed. They rule in an untroubled manner, unburdened by the needs of its citizens or even by the demands of each other. It is they who give names to Eden's inhabitants – thus bringing them into a world of language, intelligibility, and sense – and it is they who rule benignly, taking only what they need. This is sovereignty par excellence.

And yet, Poussin's painting puts this picture in a different light. In rendering the humans as part of the earth – rather than above it – Eden appears as a land given, but not taken. While Adam and Eve have dominion, it is unclear whether they have sovereignty. This is the tension at the heart of Genesis and indeed at the heart of the question of culture. Culture, like Eden, is something we have been bequeathed. A gift arriving from outside us, from a past immemorial that we had no hand in making and is not our own in any possessive sense. And yet, it is something we claim as if it were ours. Such a situation is both necessary and impossible. For Adam and Eve, the inclination to claim Eden comes with knowledge. It is only when they eat that they know that they are naked, that they are no longer under God's protection, exposed to the wind and rain of the earth. "Cursed is the ground because of you," God says, "in toil you shall eat of it all the days of your life" (Genesis 3:18). In expulsion, the first humans now know that there is an *outside* to Eden, a world where they now require shelter and where they must be concerned about soil and bread. Knowledge, sovereignty, and expulsion are thus not only intimately entwined but entwined in a manner that is paradoxical. Before exile, Adam and Eve ruled without claiming. But at the very moment they recognize Eden as their own – as something they can possess – they are expelled. Knowledge, in the Talmud, is a necessary burden. Is it necessary Levinas asks? It is necessary, he answers. Before this knowing, there was no claim. And as soon as there is a claim, there is expulsion. If the humans wanted to reclaim their rule, they would need to forsake their knowledge. But to forsake their knowledge would mean to abandon their claim. Return is not simply forbidden, it is impossible. We cannot be masters of our world and simultaneously know, possess, or self-consciously care for what we master.[2] Eden is another dream of sovereignty, another dreaming of islands.

The aim of this final section is to develop the previous discussions of dwelling, subjectivity, ontology, and sovereignty into a coherent theory of culture. To be clear, the following chapters should not be thought of as an application of these ideas – or at least not a direct application. Rather, they develop the ideas introduced in the last section into a theory; a theory that illustrates how and why subjects claim the world as their own world; a world not simply given but taken; a world where subjects dream of ownership. Two concepts from the previous section play a particularly key role. The first and most obvious is the subject. Here, the chapters carry on the work of undermining the relational and becoming model of subjectivity currently dominating geography, anthropology, and the social sciences more broadly. As already argued, the subject's essence is its vulnerability; its desperate search for sovereignty in the face of existential alterity. The second key term is that of ontology. Rather than understanding ontology as a foundational ground ordering cultural worlds, this section conceives of them as fragile responses to the problem of being a living being; representational economies built to mask (in light) a more primordial darkness. Taken together, the chapters present an alternative conception of culture, or more accurately, of cultural ontologies. Cultural ontologies are not understood as relational systems of difference. They are claims to difference. Like Adam and Eve, subjects can never properly inhabit the inheritances they have been bequeathed. While the world we are born into provides certain gifts (i.e., languages, histories, rituals, cosmologies), it never allows us to take them, to own them. As discussed in section 1, we have no say over our difference. It is given from a time out of time. But cultural ontologies provide a means for claiming ownership, building what they have been given into totems of sovereignty; representations that tell us that the world we are in is our own world. Thus, rather than approaching cultural ontologies as relational systems of difference, this section presents them as representational dreams of difference: dreams of sameness, dreams of sovereignty, dreams of ownership. The aim is to illustrate both what precisely I mean when I term cultural ontologies a representation (a dream) as well as why geography is key to making these dreams appear.

Chapter Seven

Culture as Claiming

All too readily we fall into the trap of assuming that the category words with which we differentiate ourselves from others are more than consoling illusions.

– Michael Jackson, *Lifeworlds* (2013a)

Invitations

In Paul Stoller's wonderful ethnographic memoir on sorcery among the Songhay people in Western Niger (Stoller & Olkes, 1987) , it is striking how often his interlocutors, teachers and friends articulate messages of welcome: enter, come into my house, come back tomorrow, come in and sit down, come back to the grain house, come to my compound, come into my hut, sit with me and listen. These solicitations crowd in on every page, a constant stream of directives all oriented towards hospitality. While Stoller's book is in many ways typical of the ethnographic memoir in its narrative arc (from outsider to insider, from social rejection to acceptance, and so on), what makes his project stand out is how deep his invitation into the Songhay world becomes. After witnessing a bird defecate on the anthropologist's head, his friend Djibo, a local sorcerer, declares that Paul has been "pointed out to him" for training in the ways of witchcraft (*sorko*). The invitation throws Stoller. He did not go to Niger to learn about sorcery, but now that the invitation was there, he was intrigued. It was an open door that (up until that point) he did not even recognize as a door. How could he not walk through?

While Stoller's initiation provides fascinating insight into Songay mystical traditions, at its heart lies a more intimate and vexed question: What happens when we enter another world? Vivieros de Castro's (2014) work celebrates the powerful illumination that can come when grappling with another community's ideas. But Stoller reveals the darker side of this equation. What he learns through his apprenticeship is not simply new concepts or even skills but a different way of relating to the world and sensing its existence. When Stoller accompanies Djibo to heal a local Haj, Djibo states that they need to look for the man's shadow double. When they find and release it, the power of its escape momentarily overcomes Djibo: "Did you see it?" Djibo asks Paul. "See what?" Paul responds. "Did you hear it?" Djibo asks. "Hear what?" Paul responds. "Did you feel it?" Paul just looks at him dumbly: "Djibo shook his head in disappointment … 'you look but you don't see. You touch but you don't feel. You listen but you do not hear.'" On his return to the compound, Stoller is in conflict: "I didn't understand what had occurred on the dune and I wondered whether Djibo had feigned his 'discovery': how could he see, feel and hear something which had eluded my common senses?" (Stoller & Olkes, 1987, p. 70). But to his surprise the chief – who had been struggling with consciousness and pain when he left – was up and about, laughing and thanking Djibo profusely.

This struggle with sense is a recurring theme in the book; an ongoing ambivalence between direct experience and that which Stoller *knows* can be true. It is an ontological struggle to be sure, but one that goes far beyond a concern for knowledge. In one story, Stoller reproves himself after a strange and mystical vision: "I am an anthropologist," he states, "a scientist of the human condition." In another, he struggles with his willingness to accept a powerful incantation: "Despite everything I had learned, I still could not accept that words could have such power" (Stoller & Olkes, 1987, p. 100). And finally, near the end, when Stoller puts a hex on a European businessman, he struggles with the efficacy of his spell, "I was frightened of this power as well as of the notion that sorcery was real … I needed more proof" (pp. 123–4). On the one hand, Stoller repeatedly tells himself that he needs "evidence," "this sorcery and philosophy I had been learning had to be illusion … certainly not something one might reduce to tangible data" (pp. 99–100). And yet, even as he invests in his realist epistemology, other kinds of data barge in: unsolicited dreams and visions, animal noises in the night (which leave no trace in the morning), sudden apparitions and spectres. Perhaps the most transformative moment comes when he is sent to a teacher in the far-away village of Wanzerbe. Upon arrival, the teacher ignores him and treats him coldly. Frustrated by his reception, he returns to a friend's compound only to be awoken in the night by a sense that something is in the room. When he tries to get up, he finds himself paralyzed. Helpless, he recites a defence incantation until feeling returns to his legs. The next morning, he goes to confront the teacher, but she approaches him with a smile: "Now I know that you are a man with a pure heart ... come into my house and we shall begin to learn" (p. 149). The event, Stoller states, "turned my world upside down. Before my paralysis, I knew there were scientific explanations of Songhay sorcery. After Wanzerbe, my unwavering faith in science vanished ... I had crossed an invisible threshold into the Songhay world" (p. 153).

It is precisely this question of crossing worlds that instigates this chapter. My interest, to be clear, is not on the modality of passage itself. Rather, it is what such modes of passage reveal about what cultural ontologies *are*. Thus far, I have drawn upon the concept of cultural ontologies in an ambivalent fashion. On the one hand, I argued in chapter 2 that cultural ontologies can become another way of naming difference and thus another way to "same" subjects at a particular level of identity. Yet in the previous chapter, I presented ontologies in a different light. For Levinas, ontologies are economies of representation and knowing that subjects build in order to be subjects, that is, in order to be *beings* despite the vampiric pull of alterity, the negative power that takes away all agency and capacity.[1] The aim here is to develop this idea of ontology into a more comprehensive theory of culture. A theory that presents culture not as a set of relations operating in a field of potential and possibility, but as set of responses, endeavouring to make sense of a life that perpetually eludes sense. Ontologies, in this rendering, are not simply fragile, they are illusory, perpetually undermined by that which subtracts force, undermines capacity, and reduces surety. Indeed, while Stoller's story is one that recounts his increasing competence in Songhay sorcery, the recurring theme in his text is about disorientation: "nothing that I had learned in academe had prepared me," he states, "I had become more deeply involved in things Songhay than I could have ever imagined" (Stoller & Olkes, 1987, p. 153). While this deep involvement signals new forms of learning, sensing, and understanding, it is also disorienting. Truth suddenly seems deficient. Reality appears fragile. Thus, even as Stoller's text endorses Viveiros de Castro's (2014) notion that other ontologies are enlightening, it also illustrates how they are fundamentally vulnerable. As Derrida (2000) suggests, every welcome embeds an ambivalence. To be truly hospitable is to hand over our homes to the other. But in doing so we put our ownership into question. This is Stoller's struggle. Entering the Songhay world was disorienting not because it revealed his world as fragile but because it revealed that it was not *his* in the first place. Entering the dwelling of another puts sovereignty (ours and theirs) in question, unmasking its representational light. The invitation is always a disruption;

a fundamental destabilization of what we understand to be ours (Stoller & Olkes, 1987).

The following chapter is divided into four sections. The first section describes precisely what I mean when I am talking about cultural ontologies. While the ideas here are influenced by the cultural ontologies literature in anthropology, they also draw upon the previous section to think about ontology in different terms. This leads to section 2, which confronts the work of Bruno Latour (2013) whose ideas have had a powerful influence on the cultural ontologies literature in anthropology. This section dives into some of Latour's key concepts to illustrate why they have been so influential and how they provide a coherent and compelling account for the existence of other ontological worlds. Section 3 moves on to a critique of Latour that takes aim at some of the primordial presumptions he brings to his metaphysics. Specifically, I argue that Latour grounds his conception of ontology in a metaphysics of force. In response, this section illuminates the limits of thinking ontologies in terms of what they do, produce, engender, and create. The final section brings the argument together by drawing upon the work of Michael Jackson (2005, 2013a) and what he terms *existential anthropology*. Along with Jackson, this chapter argues that ontologies are responses to existential problems; problems that are quintessentially outside ontology and beyond the capacity of any force. Thus, while cultural ontologies are no doubt constituted through relations, their origin does not reside in those relations.

The chapter concludes by arguing that culture, understood in the terms described in the previous section, is a claim. By this I mean that culture is one of the ways subjects respond to the condition of being vulnerable, that is, of living in and among (always and irresolvably) existential conditions that undermine the subject's capacity for self-ownership. Even as subjects build complex social objects (systems, cosmologies, institutions) to protect their lives, such entities never make the problem of being a living being go away. On the contrary, we are always beholden. As Jackson (2005) puts it, "Though human existence is relational – a mode of being-in-the-world – it is continually at risk … we are involved in a constant struggle to sustain and augment our being in relation to the being of others, as well as the nonbeing of the physical and material world, and the ultimate extinction of being that is death" (p. xiv). It is this *struggle* to sustain social relations that I am interested in; how alterity puts at risk not simply our biological lives but our sense of ownership and possession over the life we live, what Jackson (2013a) calls "one's basic ontological security, one's existential footing" (p. 214). In illuminating the problem of existential vulnerability, we can hopefully see how culture, far from being a set of hegemonic social structures, is a weak and feeble response. Cultural ontologies are modes of knowledge that subjects build in the face of the world and its mystery. And they are buildings that can never hold. The aim of this chapter is to make this basic point: behind every cultural ontology there is an *invitation*; a summons to lay claim over that which subjects have no claim. Cultural ontologies are not ideas, concepts or worlds, but are claims: demands to render being as something that is not simply *there* (Da) but something that is *mine*.

Cultural Ontologies

The aim of this section is to clarify what precisely I mean by the term *cultural ontologies*, a necessary clarification simply because our notions of ontology, the ontological, and the ontological turn have proliferated so widely in anthropology, geography, and beyond (Dupuis & Thorns, 1998; Mitzen, 2006; Subotić, 2016). As already discussed, ontological debates in anthropology mostly revolve around the utility of the culture concept, the dead-end of representational analysis, an interest in Indigenous philosophies and a broader exploration of ontological worlds (Holbraad & Pedersen, 2017). While geographers do not use the term cultural ontologies, there has been similar interest in Indigenous ontologies (Cameron et al., 2014; Country et al., 2015; Hunt, 2014; Ioris et al., 2019; Larsen & Johnson, 2012; Radcliffe, 2017; Wright, 2015) and specifically the role of Indigeneity in decolonizing geographical thought (Barker & Pickerill, 2020; Carter & Hollinsworth, 2017; Clement, 2019; Panelli, 2008; Sundberg, 2014). In addition, there have been ongoing debates over the last two decades around analogous ontological questions, such as, human non-human relations (Whatmore, 2006; Yusoff, 2015), affect (B. Anderson, 2011, 2016; Bissell, 2016; McCormack, 2003, 2007), the spatial implications of assemblage (Anderson & McFarlane, 2011; Braun, 2008; McFarlane, 2009), and the non-representational capacities of bodies (Bissell, 2009, 2015; Dewsbury, 2002, 2015).

While this project draws sustenance from all these debates, it is important to be clear that when I talk about cultural ontologies, I am using the term in the sense discussed in the previous section, that is, as a way of thinking about the lifeworld, the *Lebenswelt*, as Heidegger (1962) conceives it. On the one hand, this means that the term *ontology* maintains an element of its classical definition, that is, exploring the pre-established structures determining what does and what does not exist. However, rather than understanding those structures as being grounded in metaphysical truths (transcendent notions of the real, the universal, and the true), Heidegger grounds them in situational structures emerging from long-standing historical relations. It is for this reason that Heidegger talks about things *appearing* in the world rather than *existing*. The world discloses itself in a manner where things appear as appropriate, practical, and sensible – hammers appearing in a world where there are nails and quarks appearing in a world where there is astronomy. What appears to subjects ontically (within everyday experience) cannot be thought outside its ontological disclosure. Thus, subjects know the world in the sense that it is everyday and familiar (like we know a friend or we know our surroundings) and they know it in the sense that the worlds answers the queries proper to it. Ontological knowledge, in either case, is always correspondent to the world itself, that is, to what appears.

Such a conception of ontology hews very close to that of Viveiros de Castro (2003, 2014). Though one of the distinguishing features of Viveiros de Castro is that he understands these worlds as essentially non-transposable.[2] By this he means that we need to understand (or try to understand) ontological worlds within their own ontological terms. As he stresses repeatedly, there is not a single world with many "cultural" perspectives. There are separate and distinct worlds each with their own array of knowledges and appearances. The question for the anthropologist, therefore, is *not* how x native understands this or that object in relation to y native (or z anthropologist) who understands it differently. On the contrary, it is "*What is this object* within the ontological relations that constitute *this* world?"[3] For Viveiros de Castro, like Heidegger, human capacity for thought is infinitely deep, provisioning subjects with a multitude of possibilities for seeing, conceptualizing, and living. And the anthropologist's role is to try and understand not simply how the native sees an object (say a gift), but what the very concept of that object means – is it a commodity, a spirit, a contract, or something else that has never been considered or thought by its interpreter? In this sense, Viveiros de Castro's concept of cultural ontologies is one of the guiding lights orienting this project.

And yet, there are two crucial distinctions I would make between the framework proposed here and that proposed by Viveiros de Castro. First, Viveiros de Castro (2003, 2014) is uninterested in what we might call the real world – the world beyond ontology, objective reality (also see Holbraad & Pedersen, 2017). Indeed, to accept such an idea would undermine his entire project. For Viveiros de Castro, the point of cultural ontologies is that they are not epistemological, that is, they are not cultural frames or perspectives on a shared reality. On the contrary, they are realities in and of themselves; ways of knowing and sensing the world that establish their own real. Yet, I would argue that *there is* an objective world: a world that lies outside these ontologies and their distinct modalities of knowing. The problem is that this objective world is not knowable. I say this not from the perspective of a radical epistemologist who understands subjectivity as an interior trap preventing us from seeing outside our self. On the contrary, the objective world is unknowable because it defies knowledge (Graeber, 2015). What is real is the world's mystery. While I will talk more about this concept of the real in the next chapter, suffice to say for the moment that this argument is predicated upon a particular kind of realism – what I would term a negative realism. This is a realism that posits the real world not as a positive thing that can be known but as an infinite and essential negativity, a reality that works to cast doubt, undermine certainty, and rob conviction. Negative realism does not provide a basis for truth but quite the opposite. It reveals every architecture of truth (i.e., *every ontology*) to be a claim, an assertion of reality. In saying this I do not deny that ontologies appear real. But this appearance is sustained by its effectiveness. Scientific ontologies heal the sick just like Songhay ontologies curse the healthy. Both ontologies work within their frames of understanding and are effective within their terms. But neither of them is real. What is real is the infinitely problematic problem that these ontologies seek to explain; the problem of being a living being in a world that owes life nothing.

The second crucial distinction between this framework and that proposed by Viveiros de Castro (2003, 2014) is that I am not opposed to equating ontology with culture. As a number of ontologically oriented anthropologies have argued (Carrithers et al., 2010), ontologies are conceptual before they are sociological and thus cannot be mapped onto specific communities, territories, or places. The project I am laying out, however, is more comfortable with this mapping. While I take Holbraad's (2017) point that ontologies are first and foremost philosophies about what the world is, I also understand those philosophies not only as having distinct place-based legacies but also as serving a purpose beyond the high-minded pursuit of knowledge in and for itself. Indeed, I would argue that ontologies are a form of response. They have a purpose that is not simply political (e.g., see Blaser, 2014; De la Cadena, 2010) but existential in nature. Ontological concepts are how subjects respond to the world's mystery. They explain how subjects endeavour to establish some agency over a world that perpetually eludes and undermines every such gesture.

Taken together these points provide an outline of a distinct conception of cultural ontologies. Yet to develop it more fully, I want to bring it into a conversation with one of the more influential theorists informing the ontological turn in both anthropology and geography. This is the work of Bruno Latour. While Latour is by no means the only influential theorist operating in the cultural ontologies literature, comparing his conception of ontology with my own provides a more detailed picture of what I think cultural ontologies are and how they provide a distinctive ground for thinking culture.

Modes of Existence

For many anthropologists, Latour's (1987, 1988, 1993, 2013) work has provided a welcome means to think about the constitution of ontological worlds without the baggage of territorial notions of culture. While Latour no doubt presents a metaphysical conception for how worlds appear and endure, it is not one built upon primordially constituted historical regions. On the contrary, Latour's metaphysics is predicated upon a hyper-dynamic evolving system of relations that is active and unruly even as it is capable of forming enduring social composites. While it is far too ambitious to do a thorough excavation of Latour's metaphysics in this section, I will focus on four key terms pulled from Latour's oeuvre that I think constitute the key concepts that sustain an ontologically oriented anthropology: objects, transcendence, correspondence, and habit. While these key terms cannot capture the breadth and scope of Latour's work in any comprehensive sense, I think they outline both the architecture of his thought as well as why it can be considered *anthropological*. That is, they are the terms that help us see how Latour's metaphysics can be distilled into a theory of cultural ontologies or what he terms *modes of existence*.

The first and most significant term for grappling with Latour's metaphysics is his concept of objects. For Latour (2013), objects are not the inanimate furniture that furnish our world. On the contrary, an object can be anything that appears, whether it be a hammer, a baseball diamond, a human resources department, a rabbit warren, a spirit animal, a spider in its web. What matters for Latour is not the object we see but the cavalcade of inter-nested relations that a particular object represents. To see this, however, we need to understand how Latour's metaphysics is an open-ended galaxy of unfolding forces whose various constituent elements are enrolled in the formation and deformation of various conglomerations. This means that any unity that appears – whether it be animate, spiritual, geological, institutional – is the effect of some kind of *accomplishment*. It is the result of some operation, will or desire (though not necessarily human desire). In a Latourian world we can think about what the water desires as it forces its way through the crevices of a rock face, overflows the embankment of a river, or bursts the casings of a drainage pipe. The water in these examples is itself the concatenation of molecular bonds, gravitational forces, and occasional human engineering, all of which work to bring the water into certain relations that may or may not contain the forces present therein. This means that understanding any specific unity involves examining its composition, that is, illuminating the wide array of elements and forces coming together within a particular situation to engender a specific operation, phenomenon, or event. It is for this reason that Latour (1987, 1993, 2013) often talks about these unities as quasi-objects. While the lamp on my desk appears as a unity with a straightforward function, its existence is the result of a wide-range of actions and operations that are invisible behind its unified form: not only the manufacturing networks that put the lamp together, but the years of science and engineering it represents (the discovery of electricity, the creation of efficient lightbulbs, the establishment of electricity grids), the natural elements that need to be harnessed and organized (electrons running through wires), and the economy and institutions of education and learning that would demand such objects in their world. Thus, even as the lamp itself appears stable and unyielding, its existence is akin to a house of cards. It is a structure held together by forces that are (when disaggregated and examined in their disparateness) quite fragile, open to being disrupted at any point. Indeed, I have changed the lampshade on this lamp twice, the fitting once and the light bulb several times. It is an old lamp and continually falling prey to other forces, undermining its unity and functionality. The key to understanding objects, therefore, is to approach them as fragile accomplishments. Their existence as unities is predicated upon certain constellations of elements coming together long enough to function effectively.[4]

The second term I want to explore is transcendence. While the term *transcendence* has a very specific definition in traditional metaphysics, the way Latour uses it illustrates how he secularises metaphysics into a system of everyday forces (into a system of physics) rather than a system of other-worldly abstractions. The origin of transcendence in Latour (2013), is in the frail unity of objects. As suggested above, every object in Latour is a composition, a concatenation of diverse forces that come together in the course of some (physical, chemical, economic, metabolic, spiritual, institutional) process or operation. What makes these compositions frail, is the fact that every agent contributing to it is simultaneously available to other compositions. Adam S. Miller (2013) calls this the "resistant availability" of objects, "every object resists relation even as every object is available for it. No objects are wholly resistant and no objects are entirely available. Objects are constituted as such by this double-bind" (p. 39). This explains why every object is only ever a provisional unity. The actants that make up a distinct composition are available to that unity even as they are also always available to other potential relationships and thus other unities. While the flow of a river represents a certain composition, those forces can be peeled off to serve other unities, even to the extent that the flow is stemmed or transformed into a different unity, such as a beaver lake or a wadi. The point is that all actants are double sided. A river is available to be bridged, damned, dried, engineered, fished, and so on, but it can also resist all those things. The question of whether a river is available or not depends on one's perspective and the forces that perspective itself represents.

In this sense we might think about the transcendence in Latour (2013) as a directional transcendence rather than as an other-worldly transcendence (Hannah, 2019).[5] To see this let's take Heidegger's (1962) famous man in his workshop. A man is sanding wood and encounters a particularly bothersome knot. His attention is intensified even though in the back of his mind his bladder is telling him it needs emptying and he has been feeling hungry for the last hour. Then the phone rings from the house and he is pulled away from the sanding to answer it. The example illustrates the array of simultaneous networks and compositions we are enmeshed in all the time and the different ways they express their availability. Hunger and the man's bladder nag in the background but at this point, they are weak metabolic forces incapable of overwhelming his attention on the wood. But then the phone erupts as another force that breaks his attention. The house of cards that held the distinct composition (sanding the wood) is broken and other relations are formed. Higher faculties are engaged that make a plan (answer the phone, go to the toilet, make a cup of tea with some biscuits, return to the project for an hour before I make dinner for the kids). The point here is that it is the transcendence of objects (their multi-directionality) that makes any network fragile. This transcendence is not a matter of being withdrawn or unavailable. On the contrary, it is a consequence of all objects being overavailable. Objects are always in demand and always demanding. What matters are the forces behind them and the relative strength and weakness needed to forge certain compositions.

The final point to make about transcendence is it is perhaps the hinge point for transforming a metaphysics about objects and their composition to a theory about cultural and cultural ontologies. To think about objects as transcendent is to recognize that the relations through which humans constitute the world and their place within it are fundamentally weak. To say that all objects are transcendent is to say that they are also always fundamentally vulnerable, that is, fundamentally available for – and exposed to – other compositions and forces. We know that the human body (like any body) is exposed to myriad forces seeking to transform and transmute it into ecologies, habitats, and compositions more amenable to their own flourishing. We know that our food source is not by any measure *ours* but is open to others seeking to enrol it into their own operations. While humans respond to these vulnerabilities by engendering new agencies (fertilizers, pesticides, equipment) to weaken these emerging compositions, none of these networks makes the transcendent nature of objects go away. On the contrary, it is an ongoing battle. Objects by definition resist because they are transcendent, that is, always available to being enrolled, disaggregated, and decomposed through other relations.

This brings us to the third term I want to pull from Latour's metaphysis, which is his conception of habit. To understand habit in Latour we need to broaden it from the positions I laid out in chapter 4. For Latour (2013), habits are no doubt corporeal, but they are first and foremost compositional and (critically) institutional. In short, habits are concatenations of forces and actants fortified through layers of relational struts that work to keep them stable. They are an achievement of stability; a stability that is actively maintained through the realization of bringing relations into alignment. While these compositions are still under assault all the time, their alignment has led to a certain interreliance where one actant expects something from the other to make the composition work. In this sense, "the discontinuities are not forgotten, but they are temporarily omitted, which means that we remember them perfectly well, but obscurely (clearly) in a very particular sort of memory that we risk losing at anytime" (p. 267). It is this particular sort of memory that makes habits readily adaptable to new circumstance (what Latour calls felicitous and infelicitous conditions). The example Latour gives is of vultures that turn from scavenging to hunting in the wake of new European laws that require farmers to clear dead carrion from their fields. Here we have habits being open to new alignments, but it is a matter of adjustment rather than full blown recomposition.

The final point to make about habits – and the final point to make about Latour – is that they offer what he terms *paths*. These paths are essentially established institutional signposts that allow human beings to navigate (with some confidence) a hyperactive world. They are achievements: knowledges established through a long history of experimental encounters that work to create solutions to a world of myriad demands. This is the closest that Latour (2013) comes to discussing what we might have called (in another age) social structures, but not in the traditional sense. While they are architectures of knowledges and expectations, they are held together not by some underlying order, force, or system but via the simple fact that they are effective. Whether we are talking about Western science or Songhai sorcery, these are institutional modes of knowledge that subjects invest in because they lead beings to where they need to go. In this sense, Latour provides a pragmatic (rather than abstract) understanding of why and how compositions endure. Cultural ontologies are strong as long as they work, that is, as long as they are resilient and helpful in managing the various demands that summon us, providing a path that assists us in finding our way. In addition, while the variable of effectivity makes such architectures possible, it also makes them weak, constantly falling apart in the face of other demands creating disruption and/or leading towards new potential paths.

Wounded Ontologies

Before developing a critique of Latour's conception of ontology, it is worth noting the many things I like about his framework. First, I am attracted to its pragmatism, the idea that ontologies are sustained by their effectiveness, that is, by their capacity to establish helpful paths through a fluid, unpredictable, and chaotic world. Second, I am attracted to Latour's emphasis on fragility, the idea that ontologies are perpetually available to other relations and coalitions. If pragmatism is the engine of coherence in Latour, then fragility is the engine of disruption and movement. Yet, it is on this second point where I begin this critique. For Latour, what makes an ontology fragile is its inherent transcendence, the fact that (as an object) it is multiply available to many relational compositions at the same time. Ontologies, in Latour, are overdetermined – that is, always and unremittingly drawn towards other relations and movements.

In response, I draw upon Levinas to frame ontologies in a different light, not as overdetermined but underdetermined, or perhaps more accurately, negatively determined; determined not by their availability but by their failure to touch and be touched. To understand this critique, we need to recognize that Levinas and Latour operate from fundamentally different metaphysical starting points. For Latour, all action, living, and being, starts with force. Even though the emphasis of Latour's work is on examining the complexity immanent to all systems, the founding premise of this approach is that force is *apriori* – that is, it does not need to be justified, accounted for or in any way explained. Force just is. It is the primordial, predetermined dynamic animating and sustaining any and every ontological formation. Levinas, however starts from a position of primordial negativity. Life, for Levinas, is not a response to other forms of life. On the contrary, it is a response to death, to the problem of not-living. In this framing, actants and forces are not primordial but are modes of response. They emerge from the pre-ontological condition of being permanently and irreparably exposed to something outside them; something that is not simply outside a being's power to manage, manipulate, or control but also outside its capacity to protect itself, to find cover, from the problems it delivers. This is a condition that Derrida (2001) terms *woundedness*. The central argument here is that woundedness, not force, is the primordial condition for every ontology.

The notion of the wound appears in many areas of Derrida's work and not always consistently. At some points, Derrida (2005b) refers to the practice of translation and interpretation as wounding in that it necessarily injures or violates the singularity of a text. At other times, he discusses the aporatic itself as a wound in that it positions two irreconcilable possibilities at the heart of certain "ordering" of terms like *friendship* (Derrida, 1997), *democracy* (Derrida, 2005a), or the *gift* (Derrida, 1993). The notion of woundedness that I want to focus on here, while related to these dimensions, is also somewhat different. As a condition, woundedness names a limit that instigates and simultaneously mortally injures every claim to knowledge, determination, or resolution. One can see this most obviously in Derrida's (2005a) discussion of trauma after the events on 11 September 2001 (or 9/11). For Derrida (pace Freud), the definition of trauma is that it leaves a double wound. First, there is the wound of the event itself – the crash, the terror, and the death – and then there is the wound that comes after the event; the reckoning with the fact that events like 9/11 *can happen* and can happen at any time.[6] Indeed, for those of us who remember 9/11, the following weeks and months were characterized by social hysteria; a moment when society seemed perpetually braced for the next attack. Derrida (2005a) describes this pathological anticipation as a time out of time; a time when another 9/11 *could* happen, "a wound that has *not yet taken place*" (p. 105, emphasis added). It is precisely this *not yet taken place* that reveals this second wound; a wound born of an exposure to something hidden, anticipated but unknowable; an unimagined (and unimaginable) possible, invisible to all but waiting. This wound, for Derrida, is the raw abrasion of being exposed (unremittingly) to this potential;[7] an exposure that cannot be mitigated, protected against, or in any way planned for. Indeed, it is a wound precisely because it marks a threshold of pure *susceptibility*, an exposure to the world that cannot be sutured, dressed, or in any way made safe. The trauma of 9/11 did not create this wound (it is not a scar). Rather, the events of 9/11 exposed life *as wounded*. It was a trauma because (like any trauma), it revealed this threshold (this exposure to precarity) to be something that was always there. Indeed, this is the key point: the wound is *real*. The fact that it is not positive or present or provides any means for relation does not mean that it is abstract or imaginary. On the contrary, precarity is *always there*, even as it can only be defined, described, or denoted by its "not-thereness" – by not being present but always (unmitigatably) possible.

It is only when we see ontology through the lens of woundedness that we can see its fragility in a different light. As J. Butler (2005) suggests, subjects cannot argue with their woundedness. But they can respond. As complex networks of knowledge and expectation, ontologies are responses that create paths (in the Latourian sense) that human beings use to claim some sovereignty – some control – over the unpredictable world to which they are beholden. Yet this means that those ontologies themselves are – at their core – imminently *wounded*. In other words, even as ontologies are contrived to provide protection – to provide some shielding against the raw abrasion of living – they have no power to suture the wound itself. It is the wound that accounts for the fragility of the worlds we create. While other relations matter, the origin of all compositions, indeed of all modes of force, willing and acting in total, is the wound. Underneath every market system, every mode of governance, every system of knowledge, every institution designed to foster, secure, or facilitate life, there is a wound. Is it not the always present possibility of hunger that calls for global food systems, and all the scientific measurement, institutional planning, market analysis, and struggles for social justice they embed? Is it not the mysterious needs of other people that call for systems of governance (regardless of scale or political system) (Rose, 2019, 2021b)? The point is that at the heart of every ontology – at the core of every ontological system of knowing, power, and control – there lies a wound, "a hurt at the heart of living; a tear or laceration that is always open" (Joronen & Rose, 2021, p. 1405). Levinas (1981) calls this wound the "hither side of ontology," a pre-ontological dimension that gives every ontology its cause. Ontology, in this rendering, is fragile not because it is beset by other forces but because it is always compromised by the wound it attempts to suture. A wound that "can be shielded, tended and/or otherwise protected (by ourselves and others), [but] cannot be healed … on the contrary, [it] is always there, something we live with, even as it exposes the utter fragility of living" (Joronen & Rose, 2021, p. 1405).

When viewed through the lens of the wound, Latour's ontology looks similar, even as it is different. It looks similar because ontology is still a relational system that provides paths: pragmatic, institutional arrangements (like culture) that give guidance and comfort to subjects attempting to navigate a complex world. But it is different because these paths are not simply fragile – they are illusory. This is the key distinction between ontologies instigated by relations and those instigated by the wound. For the former, ontologies are fragile because they are beset by other relations. In the latter, they are fragile because they never effectively resolve the problem to which they respond. On the contrary, ontologies cover the problem over, hide it behind relations that address the wound but never diminish its vampiric power. In this sense, what ontologies provide is an appearance of effectiveness, a veneer of resolution that never fully holds.

As an example, we can look at Foucault's (1995, 2003, 2007) concept of state power. While Foucault is often read as a theorist interested in the constitution and ubiquity of state power, I have argued that he is more accurately a theorist of how states manage to make state power appear (Rose, 2014). The key to Foucault's analysis is illustrating how the state's capacity to establish certain problems of living as "problems of state," justifies the creation of various apparatus that not only work to address the problem (within the terms the state sets) but also reinforce the problem as a real problem. In the case of bio-power, for example, the state declares population to be a state problem and develops techniques of measuring, calculating and tracking to make the state appear *as if* it were solving those problems. But to what extent do these techniques actually protect the lives of its citizens? My point here is not that these techniques have no effects but rather that their effects take shape in terms already established by the state. In other words, by using these distinct "techniques of state," the state appears *as if* the problem it set were under its purview and control. This is Foucault's great insight: the key to making state power appear present is establishing the terms by which it is recognized and measured. Stoller's encounter with sorcery provides a different – though conceptually similar – example. For Stoller, sorcery could not be seen as truth because it could not be measured within the terms of his own ontology. The key for Stoller was recognizing the limits of his ontology and thereby opening himself up to other ways of sensing and seeing. In this manner, Stoller is able to recognize other kinds of ontological knowledge – other modes of knowing, measuring, and assessing what exists. It is only by recognizing other ontological possibilities that sorcery came to appear real. The point with each of these examples is that ontologies establish their own parameters of the real – that is, of what we can see. Does this mean that sorcery and the state are pure illusions? No. But as ontologies they work by covering over the wound to which they respond and shining their own distinct light on what they establish as present, abiding, and true.

As a final example, think of our modern food and health systems in the UK. These systems are predicated upon complex modes of knowing, organizing, and institutionalizing that work to deliver food and health care effectively (or effectively for some). But as was made painfully and powerfully clear during the pandemic, these systems do not make hunger and disease go away. This is what I mean when I call these paths illusory. They are not illusions (i.e., they are not magic), but they are a trick of the light. Ontologies work because they focus our attention on the paths they set. They work by illuminating a way to navigate a wounded world. But such paths also hide. They make visible what they can do and cover over what they cannot do – what they can never do. Beneath the state, the global food system, the market, and the institutional orders these terms facilitate (for knowing, measuring, calculating, organizing), hunger and disease are still *there*. And when those systems break down (which they indisputably do), the dominant classes that build them are always surprised, disrupted and terrified by the vulnerability that (up until that point) they did not see.[8] This is how ontologies work. Their seductive compositional power provides a luminous spectacle of security and possibility. But in focusing on the systems that appear, we miss the infinite darkness in which those systems sit, the soft tissue that marks our exposure to a dimension of living over which we have no power, no capacity, no prerogative. This is the wound that is the origin of every ontological order – of every desire to claim.

An Existential Anthropology

The remaining question, therefore, is how does this conception of ontology reorient our understanding of culture? How do wounds help us conceptualize culture in an original and distinctive manner? This is a key inflection point in the argument – a moment where (I hope) the ideas developed in the previous chapters come together in a manner that clarifies what I take culture to be and why I understand culture as a claim.

I want to begin with a story from the anthropologist Marilyn Strathern (1991). In her discussion of forgetting and cultural exchange among the Mountain Ok people of Papua New Guinea, she illustrates how the secrecy of sacred rituals creates an ever-present danger that traditional knowledge will be lost. Perhaps the ritual has not been practised for some time or (most often) it is simply forgotten by its practitioners. The point is that Indigenous knowledge is continually at risk, which raises the following problem: "How [do] the Ok manage to achieve a sense of completeness in a world where they are constantly aware that transmission is a hazard, that when paraphernalia is lost or an old man dies they are robbed of a capacity to act" (pp. 96–7)? This is where cultural exchange comes to the fore. It is the practice of borrowing and adapting traditions from neighbouring regions that mitigates the problem of forgetting and that sustains the coherence of sacred traditions from one generation to the next. For Strathern, this is an interesting paradox. In contrast to the romantic idea that Indigenous knowledge is rooted in autochthonous memory handed down from one generation to the next, she illustrates that such knowledge is always being *lost* and thus in need of constant replenishment. In other words, what sustains Indigenous knowledge is not its endemicness, that is, its intrinsic belongingness to a particular place. What sustains it is loss itself – the fact that things are forgotten. It is the problem of loss, Strathern argues, that "stimulates" the Ok people "to making present images work" (p. 98).

What I find interesting about Strathern's story is that forgetting is not presented as a force. Rather, it is a condition. It happens. Like hunger and illness, forgetting is an ever-present part of existence. While human subjects take many actions not to forget (we memorize, write, photograph, archive, house, back-up, etc.), such actions have no effect on the unremitting process of forgetting itself. The actions are strategies that attempt to stem the tide, plucking out chosen events and using various prosthetics to remember them (Stiegler, 1998).[9] In doing so, they reiterate ever more clearly that forgetting is the norm. Hence Strathern's (1991) admonition of anthropologists who over lament the loss or break-down of handed down knowledge, as if such breaks in transmission marked the break-down of society. On the contrary, what Strathern's discussion illustrates so vividly is that loss is the instigation of cultural solutions. The constitution of knowledge (and the actions people take to build that knowledge) is anchored in loss.

For several decades, the anthropologist Michael Jackson (2005, 2009, 2013a) has developed this insight into a broader theory of culture. Existential anthropology, as he calls it, begins with a recognition that culture is too often thought of purely in its sociological dimension, that is, in terms of the social order: how society is constructed, reproduced, and represented through various cultural institutions (norms and expectations, taboos and initiations, representations and performances, etc.). For Jackson, such approaches ignore how social life is predicated upon certain facts of existence: raising children, mourning the dead, healing the sick, suffering, and enduring together. These conditions work to both engender social relations and dissolve them. Thus Jackson (2013a) concludes that "a strictly sociological perspective had to be complemented with an existential perspective that encompassed the role of contingency, playfulness, unpredictability, mystery, and emotion in human life as well as acknowledging that human beings are motivated not only by a desire to construct social worlds in which they can find a sense of security, solidarity, belonging, recognition, and love but by a desire to *possess* a sense of themselves as actors and initiators" (p. 14). In this framing, culture is thought of as an active creation; what Strathern (1991) might call the practice of "making do" in order to possess a sense of consistency in one's self and one's world. Jackson's approach can be laid out through three basic propositions: (1) that culture has an existential dimension, (2) that subjectivity is claiming some agency in the face of this dimension, and (3) that culture provides a means – what Latour might call a path – to facilitate these claims.[10] I will explore each of these in turn.

In terms of the first proposition, Jackson (2013a) argues that anthropological interest in sociological structures have made them blind to the exigencies that impinge upon human life; demands, and problems whose origin lies outside culture, society, and ontology and yet are immanent to life itself. Jackson refers to these demands as extrahuman because they reside outside the social and defy every human ambition to bring them under social domination: "The world of the extrahuman contradicts our anthropomorphic assumptions and proves refractory to our intersubjective strategies of constructive engagement and control. It may be comforting to believe that we can negotiate reciprocal agreements with enemies, appease the unruly elements with sacrificial gifts, or strike bargains with the gods, but in reality the extrahuman world impacts upon us in unpredictable and ungovernable ways" (p. 194). The extrahuman exists (if that term can be used) not as something *there* (a present and threatening force) but as something that *isn't there*; not the possible, but the impossible that we cannot see: the mystery of the future, the inescapable frailty of our bodies, the dark interior of another human being, always elusive and infinitely surprising. These unseeabilities haunt human life at the limits of what we can know, understand, and predict, much less control, manage, or account for. And yet the fact that we cannot touch them does not mean they cannot touch us. On the contrary, they touch us all the time, upending our plans and lacerating our lives. As Jackson states, "In the modern world, we have become so used to the idea that science can save us from an untimely death, and that the state can protect us from threats of invasion ... that we expect, almost as a constitutional right, a painless life … but suffering is … an inevitable and unavoidable part of life; it will always have to be endured" (p. 226).

Indeed, suffering is perhaps the most effective way of understanding the remorseless presence non-presence of the extrahuman, our primordial woundedness. There is the suffering of our bodies through injury and there is the suffering of our heart through grief and heartbreak. But there is also a more generalized suffering that Jackson (2013a) is illuminating; a suffering that comes from being beholden to others over which we have no say. These others are human others (who cannot account for the demands, desires, and pain they impose) but also non-human others (mudslides, weather, overgrowths, and extinctions). The point is that Jackson's existential anthropology is founded on the idea that the extrahuman is the underlying engine of cultural forms: "Underlying historically determined forms and culturally approved practices lie recurring existential needs and conflicts" (p. 216). For this reason, he shifts his focus from culture itself to the problems it seeks to solve: "An existential focus on situations helps us see that people are not simply determined by dominant epistemes … but opportunistically engaged in a search for technics whereby life can be made viable within the limits set" (p. 276).

This notion leads to Jackson's (2013a) second proposition, which is that human subjectivity is the project of attempting to claim sovereignty. To illustrate the point, Jackson refers to Freud's discussion of his nephew's game of mastery in his 1955 essay, "Beyond the Pleasure Principal," what Freud called the *fort da*. The game was invented by his eighteen-month-old nephew to cope with his mother temporarily leaving him on his own. In the game the boy threw a small toy out of his cot so it was out of sight, at which point he said *fort* (gone). The toy, however, was attached to a piece of string and the boy would use the string to pull the toy back at which point he would exclaim *da* (there). For Freud, the game illustrates how the boy comported himself to his situation. His mother's leaving was not something he could control but something he needed to bear. But he could create an illusion of mastery; a modality of play whereby he could summon his mother's return (*da*) in a fantastical way. As Freud (1955) states, "At the outset he was in a passive situation – he was overpowered by the experience; but, by repeating it, unpleasurable though it was, as a game, he took an active part. These efforts might be put down to an instinct for mastery" (p. 16).

It is this instinct for mastery, this inclination to turn our suffering into an illusion of control, that intrigues Jackson (2013a). As he explains, the human condition is one that works to transform itself from being an object, exposed to the vicissitudes to the extrahuman, to being a subject, someone who has agency (even if it is illusory) over their life and its direction: "Nursing ill will toward an enemy [or] cursing an errant computer ... will not necessarily effect any change ... but it may reverse one's experience ... one becomes, imaginatively and retrospectively, the determining subject of the events that reduced one to the status of an object" (p. 198). To illustrate this point, Jackson (2013b) describes storytelling as a form of sorcery: "We deploy words magically to ... give the impression that we actually grasp the hidden meaning of the world in which we move ... an arcane technique for consoling lost souls that the world was indeed within their grasp" (p. 37). A wonderful illustration of this point is made by the anthropologist Cheryl Mattingly in her study of occupational therapists. As Mattingly (1998) notes, the experience of healing seems to require finding a story or plot that will move a patient from a condition of being passively beholden to their condition, to one of (if not overcoming it) learning to live in a manner where they (once again) are the central protagonist. By turning each milestone into an episode in an unfolding narrative, the story "offers a way to examine clinical life as a series of existential negotiations between clinicians and patients, ones which concern the meaning of ... a life which must be remade in the face of serious illness" (pp. 20–1). Storytelling, in this rendering, is a form of master play, a way of narrativising life in a manner that "enables us to overcome a sense of being existentially diminished" (Jackson, 2013b, p. 16). Are such narratives necessary? They are necessary. While master play is a modality of play (in the sense that the control it narrates is illusory), it is a necessary mode of play (in the sense that it gives subjects a means to claim sovereignty over their world). As Geertz (1973) might suggest, master play is the stories we tell ourselves about ourselves. And while those stories have no bearing on the existential reality of our woundedness, they allow us to bear that condition through imaginary "plots."

This leads to the final point that emerges from Jackson's (2013a) work. Master play is not the result of spontaneous creativity or individual invention. On the contrary, it emerges through well-worn paths; institutions set down by others whose function is to provide what Jackson calls *ontological security*: "one's existential footing" (p. 241). For Jackson, handed-down technics are bequeathed to give subjects tools to navigate an unreliable world. While these technics are illusory, they nonetheless work to establish a veneer of rules, protocols, and practices that tell us that the world is in hand. Such a conception of culture stands in stark contrast to theories that present culture as something imposed on otherwise "free" subjects. Once we start with a subject that is primordially wounded, culture comes to appear as a refuge; an inheritance subjects embrace and build because it offers resources for understanding oneself as a subject – that is, as a being who is sovereign. This is what Jackson means by ontological security – not the security of a foundation but the security of knowing that one has means (tools, techniques, paths) that one can use to exist. While these means are illusory, they are nonetheless there, in the dwelling, furnishing subjects the means to define themselves in the face of the extrahuman.

It is at this point that we can see most clearly what I mean when I argue that culture is a claim. My argument is that cultural ontologies are not simply ideas about the world. While the ontological anthropology literature illustrates how cultural ontologies frame what other people think, Jackson (2013a, 2013b) illustrates how they frame what all people need. It is only when ontologies are understood existentially, that is, as responses to the condition of woundedness, that we can see how human experience is not governed by forces and relations but what might be thought of as their opposite: a lack of force, a seeping away of capacity, the ever-present non-presence of existential alterity. Seen in this light, ontologies are beacons. They are inherited paths that subjects cultivate in order to claim some agency – some sovereignty – over a world that perpetually eludes them. Indeed, from here on out one will notice an easy slippage between the terms *culture*, *ontology*, and *cultural ontology*. This is because ontologies are understood as paths – inherited ways of knowing. But the fact that these paths are given does not explain why they are taken. This is why they are not simply ontologies but *cultural ontologies*. They are modes of knowing that are claimed. Recall that this was the opening question of the book: Why do people claim their worlds? The aim of this chapter has been to illustrate how woundedness is the engine of every cultural ontology; a system of master play built (via well-worn paths) to engender an illusory sense of security.

Conclusion

In conclusion, this chapter has outlined a conception of culture as claiming. While it begins with Latour's (2013) notion of ontology as a "mode of existence," it situates that ontology within a broader set of existential conditions. In doing so, cultural ontologies appear in a different light. Rather than being fragile systems of relations and knowing, they appear as incomplete responses to the problem of not-knowing and the ever-present non-presence of the wound. The difference is subtle. For Latour ontologies are fragile – always exposed to the disruption of other forces. In the framework I am proposing, ontologies are wounded, desperate, and always incomplete attempts to secure themselves against alterity. As systems built in response to the wound, they have no substantive hold over the problems and conditions they seek to resolve. And yet we live in hope, building a world that appears *as if* it were predictable, calculable, secure; a world where we are *subjects,* that is, active agents in the securing of our own lives. This is a subject that is not determined by its relations but by its sense of propriety, that is, by its sense of being something more than a sail in the wind, tossed by current and sea.

Understanding culture as a response to the wound no doubt carries certain dangers. Most obviously, the framework has the potential to be misread as a return to the one nature, many cultures framework, whereby culture is understood as an overlay to a more primary real world. While I do not deny that my argument moots a particular conception of what is real, this real is not a real world. Indeed, this is the paradoxical point of alterity: it never properly appears; it is never present; it never *is*. The wound is there only in its not-thereness, an absence we cannot see even as it solicits various ontological responses. Another concern is that such a conception of culture makes cultural ontologies appear somewhat uninteresting if not prosaic. This is perhaps a fair critique. While I am not arguing that difference itself is uninteresting, I am querying the oft-cited position that we need to take cultural ontologies *seriously*. Ontologies are not worlds. They are claims to worlds. And while I do think we need to take these claims seriously, it is equally important not to be seduced by the worlds they claim. Thus, even as this work is influenced by ontological anthropology, my interest lies less in the ontological concepts themselves, and more in the way they respond to the extrahuman. Along with Jackson (2013a), I see difference as essentially ordinary: "alternative ways of addressing recurring universal questions of existential viability" (p. 26). The world *is* ontologically determined. But those determinations are weak. And we should be wary of taking them too seriously. This becomes particularly important when we consider the various ways ontologies represent the world. But this is the topic of the next chapter.

Chapter Eight

Fundamental Geography

Do we continually have to prove to ourselves that we exist? A strange sign of weakness ... endlessly self-evident.

– Baudrillard, *America* (1989)

Introduction

I want to open this chapter with one last story from Michael Jackson (2013a). Here, the anthropologist is travelling with an Aboriginal companion named Zack, a senior custodian of the sacred site where he is camping. After incurring a flat tire on the road, Zack insists that the local mining company is obligated to fix their jeep as the company operates within Aboriginal territory. When they come to the padlocked gate separating the mining compound from the rest of the territory, Zack takes the gate off its hinges and lets the jeep pass through, proclaiming "this is our country" and "we are boss here." But when Jackson and Zack drive into the compound itself, this confidence begins to fade. No longer are he and Zack "in country," camping amidst the Dreaming sites that Zack maintains. Now they were in a typical Australian mining settlement complete with modular offices, planted gardens, and air conditioning. When they arrive at the main office, Jackson states that Zack is clearly nervous: "In his bare feet, frayed bell-bottoms, and grubby windcheater, Zack looked as incongruous as I felt ... his manner was an odd mixture of bravado and deference. After forthrightly announcing that he was a traditional owner of the Granites [mine], he stood cap in hand, so to speak, waiting for the mine manager to respond" (p. 240). In Jackson's analysis of the scene he argues that the event represents a limit experience for Zack. Here, among the linoleum and fluorescents, Zack's sense of ownership ebbs away. At the boundary of his world and that of another, Zack loses his composure; he is no longer self-possessed.

Geographers have known for a long time that landscapes and places communicate to us ideas about who we are and what kind of world we belong to. In exploring our everyday streets, sidewalks, playgrounds, and monuments, they are highly attuned to how places inculcate and reinforce a range of powerful social and cultural ideas about what we take to be normal and proper. And yet, they have also had a tendency to think this relation – between identity and place, culture and landscape – in somewhat narrow terms. In the traditional work of new cultural geography, for example, identity is conceptualized as something located *inside* the subject, part and parcel of their being. Geography, therefore, is what is outside. It expresses an identity already presumed to be interior. Indeed, the very idea of the social and cultural is predicated upon this relation, the idea that identity is within us, waiting to be socialized, acculturated, dominated, classed, or raced. Identity is the cypher – the key to the subject's being – and the landscape (and landscape struggle) is one of the means by which this key can be turned.

Given this formulation, we might expect Zack's discomfort to be a result of the mismatch between Zack's internal identity and his external environment. The dominant landscape of the mining settlement does not represent him or acknowledge his identity and therefore Zack feels out of place, excluded, and unwelcome. But the issue, as Jackson (2013a) describes it, is not that Zack feels somehow malfitted to the world around him. He feels diminished, incapable, lost. In other words, the problem is not that the material world is at odds with Zack's internal identity. The problem is that Zack's claim to an internal identity becomes unmoored, dislocated, set adrift. It is among the rocks and gullies of "the country" that Zack feels at home. This is where Zack can see his place in the world, his sense of self supported by its very materiality all around him. But among the coffee machines and photocopiers, such signs are erased. His sense of self lost without its visible anchors. This is a very different way of understanding the relationship between identity and place, and a very different framework for approaching representation.

The central proposition of this chapter is as follows: What if we understand the landscape not as a site that reflects or expresses cultural identity but as a site that allow identity to transpire?[1] In the previous chapter, I argued that cultural ontologies are a form of master play: stories subjects tell themselves about themselves to narrate the idea that they are subjects – that is, that they exist not as objects of the world but as subjects in the world, capable of shaping their life and destiny. In addition, I argued that these stories are illusory, perpetually undermined by the woundedness of being a living being. And yet culture does not *appear* fragile. In the opening chapter I argued that culture appears as a violent and present force, driving political contestation and social division. How can culture not be thought of as something within us, shaping our understandings and actions? The aim of this chapter is to illustrate how the cultural ontologies that subjects (failingly and desperately) cultivate come to appear as present and abiding. To see this we need to reiterate that while cultural ontologies have no capacity to mitigate the reality of existential woundedness, they do have the ability to hide it – they outshine the absence of alterity by circumscribing a domain of light. A domain I described in chapter 6 as representational. The aim of this chapter is to elucidate a particular approach to representation and illustrate why geography plays a distinctive role in its constitution. Indeed, the argument here is that cultural ontologies are not simply claims but representational claims. By this I mean that they are things *built*. It is only by writing claims into the world – by materially representing them in the landscape – that culture comes to appear as real. This is geography's unique power; its capacity to make claims visible, present, and everyday. Geography is what gives claims substance and quiddity. It is thus fundamental to culture's phenomenality. Geography is what allows culture to appear.

The argument is divided into four sections. The first two sections focus on reconceptualizing what representation is and how representation works. Section 1 draws together Levinas's notion of dwelling (discussed in chapter 6) with Lacan's conception of the symbolic to reframe representation in ontological terms. Specifically, it argues that the cultural ontologies discussed in the previous chapter need to be thought of as both practical and symbolic. While they no doubt provide systems for knowing and understanding (a means for finding a path), they also tell a story about who we are and how we narrate our being. Section 2 develops this framework further by providing three definitions for how to understand representation. Representation is (1) a response, (2) an investment, and (3) an illusion. Taken together these definitions present representation not as a powerful act but as a precarious image sustained through representational practices that outshine the reality of alterity, or what I will come to call (drawing upon Lacan) the Real.

Section 3 explores the implications of this position for cultural geography. Specifically, it turns traditional conceptions of cultural geography on their head. Rather than the material world being an expression of an internalized identity, landscape is understood as an important means – indeed, the *fundamental means* – by which those claims are made. The final section thinks through what it means to approach geography as something that generates (rather than reflects) culture. Specifically, it argues that geography needs to be understood as a fabrication – something made and something that deceives.[2] Once we see geography in such terms, we can recognize the problem with being seduced by the landscape's materiality; seduced into thinking that its physical presence truly expresses the power and capacity it claims. Geography is fundamental because it provides subjects the capacity to claim themselves in material terms. But we need to be wary of such claims and not be seduced by the material forms through which they speak.

The Representational and the Real

I want to go back to a discussion I started in chapter 4 regarding Eduardo Kohn's book *How Forests Think* (2013). One of the things I love about this book is its distinctive semiotic approach. For Kohn, the forest is a system of signs. "Life is intrinsically semiotic," Kohn states (p. 76). The forest lives and thinks in the sense that it is organized around semiotic events that communicate to other beings information that they use to make decisions. Thus trees "decide" to produce a certain kind of sap in response to a particular infestation, or anteaters "decide" to gravitate to a particular tree because it is infested. These choices may not be conscious choices in the way we normally think about them, but they are responses made in relation to the interpretation of signs. The forest thinks, because it is a system for reading and acknowledging signs. And yet, I also argued that Kohn does not acknowledge enough those modes of sensing that are not instigated by positive forces; modes of experience, I would suggest, that humans are uniquely adapted to sense. Kohn's anxiety about the mud slide is a distinctive example. For Kohn, his response is indicative of a certain breakage between thought and matter; between the anxieties that circulated in his head and the material relations in which he was situated. At one level, this is no doubt true. Kohn was disconnected from the materialities of his situation. But this is because the situation offered no capacity for connection. Kohn *could not know* what the mud would do. There was nothing there for him to read, understand, or connect to. But this not-knowing is still a form of sensing. What Kohn sensed was not something present but something absent, a dark spot of information and knowledge. And it is this absence that marks Kohn's transition from thinking instigated by signs to thinking instigated by silence – silence not as a sign (as in the absence of noise) but as an emptiness, a nothing.

The story illuminates a blindness in Kohn's (2013) conception of the relation between semiotics and knowledge. Yes, all beings read signs and many beings are anxious about the signs that they cannot see. But what distinguishes the human being is not simply its awareness of what it does not know, but the fact that this awareness (or this lack of awareness) is *a problem*. This is a problem that is nicely illustrated by Zupančič's (2017) definition of sex. Sexuality, as she understands it, is an archaic mixture of desires, instincts, pleasures, and needs that are outside of representation – unconscious, unknown, and unknowable. While we as subjects may know that we desire specific things (we are drawn to specific objects of desire), desire itself (that which makes something desirable) defies knowledge. And it is precisely this defiance of knowledge that makes sex a problem for us. While I am not an expert on animal cognition, it does not seem that animals are very concerned with their desire – how it is directed, the fact that it comes and goes, that it gets attached to particular things – nor do they seem to approach it as something that needs to be *understood*. But humans are concerned with their desire.[3] And this concern endows humans with a problem. As something that escapes our faculties for knowledge, understanding, and organization, it situates a question; a problem that Zupančič (pace Lacan) understands as the origin of representation.

So what is this connection between desire and representation? In chapter 6, I argued that Levinas's (1969) notion of dwelling needs to be understood as a domain of representation; a place of rest and respite where the subject finds (Eden-like) an inherited representational economy that it uses to narrate itself as a sovereign being (claiming). This conception of representation, however, is not very well developed. What precisely does Levinas mean by a representational economy? How is it different from other economies and (perhaps more saliently) how does it differ from other conceptions of representation? These questions, while peripheral for Levinas, are central for Lacan. To be sure, Levinas and Lacan are two very different thinkers. They emerge from different conceptual traditions, work towards different theoretical ends, and (critically) develop distinct philosophical positions (Critchley, 1998; Ruti, 2015). But one cannot help noticing how their work hews to a similar theoretical arc.[4] Both thinkers anchor human life in an infinite negativity (the transcendent Other for Levinas and the unconscious for Lacan) and they both understand this negativity as the instigation of a representational economy. To illustrate how this relation is developed in

Lacan, and why this development is particularly useful, we need to better understand his conception of desire, particularly as it takes shape through his notion of the Thing (*Das Ding*).

Normally, when we think of desire, we associate it with an object, something that will satisfy or appease a perceived need. But for Lacan (2008), desire has no primordial object – not in the present (in terms of some overarching desire – e.g., genital sexual desire) or in the past (in terms of some originary relationship – e.g., with the mother). Rather, desire exists as something archaic and unformed, defying thought and description. What desire desires is something. But what that something is can neither be named nor contained by any particular object. It is not that desire is nothing. It is more accurately a "something missing," an unpinpointable absence that is sensed and felt but cannot be named or satisfied through the various objects we focus on.[5] This is the distinction between desire itself (which is shapeless and unsatisfiable) and the desired object (which is distinctive and nounal). Because the subject cannot identify or name the former, it seeks satisfaction through the latter. But the latter are simply substitutions; they objectify desire by identifying it within what Lacan calls the onto-symbolic order; the representational economy where social life takes place. Thus, while the onto-symbolic order is necessary, since it gives desire an object, the desire itself transcends that order and the objects it offers. Behind every desired object – every desired thing – there exists the shadow of something else; a desired thing behind the thing; a desire that cannot be consolidated or contained by its various representations. In addition, it is this thing behind the thing – this primary desire which cannot be seen but which shows itself through various substitutions, that Lacan (2008) calls the Thing (*Das Ding*): "This Thing will always be represented by emptiness," Lacan states, "precisely because it cannot be represented by anything else – or, more exactly, because it can only be represented by *something else*" (p. 160, my emphasis).

There are a number of implications I want to develop for thinking about representation in light of the Thing. First, like Levinas's (1981) conception of the saying, Lacan posits desire as the *instigator* of representation. The Thing provokes a multitude of representational substitutions (saids) to capture or satisfy something that escapes naming (sayings). It is precisely because the Thing defies satisfaction that subjects initiate and pursue various substitutional strategies. Thus, even though I cannot satisfy my desire, I can seek other kinds of pleasures that do not resolve, *but can echo*, this need.

Second, this means that the Thing is generative of a whole symbolic field. If we understand subjectivity as an endeavour to discharge or satisfy the Thing, then we must also understand the representational objects we find as also anchored in this same drive. In other words, the subject comes to understand and make sense of its desire by identifying it in the objects it finds already within its inherited world; a world that provides them with various substitutions that it uses to partially satisfy. This means that desire (and the pursuit of its satisfaction) is the mechanism by which the subject becomes attached to the world it finds. A world that presents itself in representational terms, with a series of substitutional objects already there. Before this attachment there is only desire – amorphous, unnameable, and incomprehensible. But once the subject takes up a place within and among the representational system it finds (the onto-symbolic order) its desire has direction, purpose, and understanding (or at least understanding in terms of the substitutions it finds). To sum up: it is the Thing that puts the mind to work. As Wine (2018) suggests, desire "is the organiser for the psychical apparatus" (p. 44). This means that the subject's search for a means to discharge its desire constitutes a foundational hypostatic event. The subject collects itself (defines itself, understands itself) by seeking substitutional strategies to give its desire meaning, direction, and purpose. And it finds these things in the world around it, with representational paths already laid out. The fact that these paths always fail to fulfil is not the point. For Lacan (2006), the subject would not be a subject at all if these paths were not there and the subject did not take-up its place among them. Thus, in a fashion analogous to Levinas, Lacan argues that the subject only becomes a subject by participating in the symbolic order – that is, by participating in the language that gives us the capacity to speak and want even though the words and objects we find never capture or fulfil. Being a subject means cultivating, caring for, and claiming the symbolic order that allows us to be, that is, that provides paths towards various (partial) satisfactions.

This leads to the final implication of this framework, which is that such a conception of representation reorients what we understand representation to be and why I equate the symbolic order with the ontological. It used to be the case (in cultural geography and Cultural Studies in particular) that representation was something self-standing subjects did to represent and misrepresent the real world.[6] What was real was objective and what was representational was epistemic. For Lacan (2008), however, "the Real" is what sits *outside* language. The Real is the domain of desire, the elemental, the saying, the Other and the Thing. While each of these concepts illuminates a unique and non-overlapping dimension of existential alterity, they collectively signal a dimension that comes before ontology and thus is more real than ontology. Thus, what is real is existential alterity and what is ontological is representation.[7] Or, to put it another way, the representational world is ontological but is not Real. "There is only being in the symbolic," Zupančič (2017) states, "*except that there is the Real*" (p. 43, original emphasis). By this she means that one can only "be" a nameable being within representation. But there is also a dimension beyond being; a dimension that is not present and has no existence per se, but is real. The manoeuvre I am making here is very similar to the one I made in the previous chapter, where ontology is no longer a ground for knowledge, power, or becoming but is rather a representational skin – what Critchley (1998) calls a protective veil – hiding the Real in its light but without ever fully outshining its darkness.

While the previous chapter emphasized the existential basis for every ontological order, this chapter illustrates how that order manifests representationally. Every cultural ontology is thus a symbolic order – an onto-symbolic order (not just cultural, not just ontological but also symbolic). Thus, even as I agree (pace Latour) that ontology establishes distinctive paths for knowing and navigating the world, the point I am making here is that these modes of knowing are *representations*. The practice of science (like the practice of sorcery) is an ontology because it provides a distinctive set of languages and practices for navigating a mysterious world. But such practices are still representational. They work by closing out the darkness of the Real and shining a light down particular paths. But in doing so they also "silenced … the stars and planets" (Lacan, 1988, quoted in Brody 1998, p. 57), overwriting their infinite mystery with the light of what science can do. It is only by walking these paths, participating in their light and the various objects they illuminate, that subjects can be subjects at all.

Response, Investment, Illusory

So what are the implications of understanding ontology in the representational terms outlined above? As already suggested, it means recognizing that representation is not something subjects *do*. For a long time we have understood representation as an intentional act – something subjects perform in order to achieve this or that political end. Whether it be to strategically position oneself vis-à-vis relations of power or to impose a dominant ideology on the landscape, representation was understood as an epistemic toolbox available for subjects to tactically play, enact, or impose their identity. The framework presented here is very different. As I have begun to outline above, representation is how subjects struggle to *be*. If we accept that subjectivity is a claim, then we need to understand those claims as representational. By this I mean that they are made using the symbolic order they find. In order to elaborate this conception of representation, this section outlines what I take to be three key dimensions. Representation is (1) a response, (2) an investment, and (3) an illusion.

In saying that representation is a response, I mean to draw attention to how representation arises from what Zupančič (2017) (pace Lacan) calls *the Real*. Thus far I have discussed the Real through many of its various manifestations: the precarity of being a living being, the arrival of the other person, the impossibility of our desire, and so on. What holds these manifestations together – what makes them collectively Real – is the fact that they defy relation. They stand for a dimension of human experience that is irresolvably outside us. And yet, I have also reiterated that they can be experienced. These exteriorities still elicit *a response*. Not a relational response – which comes from two positive forces connecting – but a relation of non-relation (Harrison, 2007a). The remaining question, therefore, is what makes this response representational? To see this I want to look back at arguments I made concerning the dwelling (chapter 6) and ontology (chapter 7). Both concepts are deployed to understand how human beings respond to the problem of the Real. And in both cases the response is understood as a form of boundary making. As Jacque-Alain Miller (2019) suggests, this boundary is a shield, "the barrier that the real opposes" (p. 27). As something that protects and opposes, the boundary allows being to take shape. It is what facilitates one's claim to being an "I." But I also emphasized that these boundaries are illusory. This formulation was most obvious in the conceptualization of dwelling as something both open and closed. The dwelling (which gives the "I" its home) is always open to exteriority since it is the reality of exteriority (what I am now calling the Real) that engenders its instigation. But the dwelling must also be closed – it must draw a boundary – because it is only through such boundary making that one's I'ness can take shape. The concept of representation that I am devising here sits in the same paradox. Subjects respond to exteriority by drawing a boundary. Yet, while the event of boundary drawing allows subjectivity to take place, the boundary itself cannot be real. The boundary must *be* open. But it also must *appear* closed. This is what makes the boundary a representation. The boundary responds to the real. But it is a representational response.

This leads us to the second characteristic of representation: representation as investment. In the traditional Cultural Studies framework, representation is an imposition; a negative force that impinges upon subjects in order to reinforce various hegemonic norms. In the framework

I am presenting here, however, representation is something cultivated and nurtured. This is the fundamental difference between a sociological and an existential perspective. In the former, subjects are presumed sovereign and culture is imposed. In the latter, subjects are presumed vulnerable and culture is how sovereignty is claimed. Representation is a resource that subjects cultivate in order to be. What matters, therefore, is not what representation means but what it does – how it generates tools to claim oneself. In this sense, subjects are not primarily invested in policing potential interpretations of various symbolic acts or events; what matters is that the onto-symbolic order is *in circulation*, available for appropriation. One does not need to know what it means to be a Jew in order to identify with it. What matters is that the representational category "Jew" is available to be said. In saying this, I do not mean to suggest that the category term "Jew" is not something Jewish and non-Jewish subjects struggle over (they very much do). My point, rather, is that the very practice of debating, defining, castigating, pillorying, fearmongering, and celebrating the category "Jew" facilitates its existence as a representation, that is, as a referent within the onto-symbolic order. Thus, while there is a long history of approaching these debates as identity struggles (which they may indeed be), they also serve a larger purpose. The struggle itself constitutes an onto-symbolic order. It is precisely the elaboration of naming practices – the continued extension of the Jewish question into wider areas of discussion and debate – that secures "the Jew" as a thing, that is, as a referent within the onto-symbolic order.[8]

This leads to the final, and perhaps most important, characteristic of representation: representation is illusory. I have already used this term quite a bit in this book and often without much caution. To say that the onto-symbolic order is illusory is not to say that it is a mirage or that it has no substance or materiality. On the contrary, the illusory world I am outlining here needs to be thought as a domain of light. The representational economy designates a site of illumination where certain ontological truths and realities can be seen. But this site of illumination shines only in contrast to the darkness it hides. While the paths we see provide practical solutions to real problems, they do not quiet the power of the Real. On the contrary, the Real is always still there – on the hither side of ontology – overwritten but not forgotten. As an example, let's look at Ben Anderson's (2010) work on emergency planning systems (also see Adey and Anderson, 2011). On the one hand, the institutions he explores represent a highly rarefied onto-symbolic domain for predicting, anticipating, and planning for the future. By creating potential representations of the future, these institutions develop stratified response procedures that appear as a sophisticated, multivalent domain of calculation, knowledge, and action. And yet, what can these institutions really do about the future? While the structures of calculation and planning are no doubt highly refined, none of these procedures for anticipating and planning for the future have any effect on the future itself – they have no bearing on what can potentially come. Paul Greengrass's 2006 film *United 93* fascinatingly captures this blindness. While it shows the immense institutional sophistication of the US Federal Aviation Administration, the military, local first responders, and federal and state governments, it also illustrates how the unfolding events of 9/11 utterly upended their normal parameters for interpreting and responding to events. Based upon real-life interviews with key figures in these organizations, the film captures how incredibly well-tuned institutions can be cast into total confusion by the unforeseeable. Thus, while onto-symbolic orders of emergency planning present us with a world that is in-hand, from the perspective of the Real such representations appear in a different light; not as powerful agencies but as hopeless gestures, groping in the dark, not fires in the clearing but matches in the cave.

This is the most important point about understanding representation as illusory. It establishes a domain of light in the midst of an overwhelming and perpetual darkness. By way of analogy, think about the host of your favourite BBC nature show marvelling at some feature of a highly evolved animal, a beast's perfectly adapted snout or a highly attuned ear. The host draws attention to this feature, waxing lyrical about its capacities and all the things it can do. But what is left out is the long history of failures that this creature left behind; the mountains of sea-floor and ash of also-rans. Evolution is a process of trial and error through death; a genocidal design whereby every adaptive success is preceded by millennia of incomprehensible loss. By standing within the light of our onto-symbolic world, our host puts this creature before us alive, present, and perfectly adapted, without reference to the dark and bitter process that gave us its current form. Indeed, in keeping with the show's own representational logic, the presenter may even attribute this success to the creature itself – for example, "this creature has (amazingly, ruthlessly, carefully, skilfully) adapted ..." – as if surviving is something it has done (Robinson Crusoe like) through its own will and ingenuity. This is the representational economy's great conceit; its ability to cover over the Real with its own light. Rather than understanding representational claims as claims – as desperate ambitions voiced in the dark – we accept them on the representational terms before us: marvels and triumphs rather than bitter accidents shaped by circumstance and death.

Appearances that Matter

Thus far, the emphasis of this chapter has been on developing a distinctive approach to the question of representation. The question that needs to be addressed now is how does this approach change our understanding of geography or, to put it another way, how does understanding representation in these terms make geography fundamental? One of the arguments I made in chapter 6 is that every dwelling needs to be thought of as something built. The terminology of building was borrowed from Heidegger (1971a) and I used it to refer to the full spectrum of representational practices that subjects use to care for their world. While initially I made no distinction between aesthetic (i.e., poetic) and material practices, the terminology of building was strategic. This is to say that I think there is a difference between representations that are rhetorical and those that transpire in bricks and mortar. As I have argued repeatedly, appearance *matters*. What is real is hidden. The world we believe in – that we cultivate and invest in – is the world that appears. Thus, geography is not one representational mechanism among others. It is the key mechanism, *the fundamental mechanism*, by which the onto-symbolic order is made present. Claims, in short, need a place. They require materiality – a sensible objective form – in order to work their miraculous light. It is geography, therefore, that makes the onto-symbolic order a present and visible feature of everyday social life. Geography does not reflect our onto-symbolic culture – it engenders it. It is what allows culture to transpire as a thing.

I want to illustrate this position through two examples. The first is Žižek's (2002) analysis of the Kachin spirit masks used by the Indigenous peoples of the American Southwest. The story he relates is drawn from *Sun Chief*, the autobiography of the early twentieth century Hopi chief Don Talayesva (1963). In the passage Žižek (2002) cites, Talayesva explains the role of the masks in the Kachin ritual. Each year the masks are displayed in frightening dances performed for the children. To the children, the masked dancers are terrible spirits that threaten to eat them unless they are placated with offerings. The key to Žižek's story comes when the children mature and learn the truth of the Kachin. The spirits that danced before them as children are revealed to be their own fathers and uncles. At one level, Talayesva describes the revelation as a great shock. Not only are the masked figures not gods but they are his own relations, thus putting his whole conception of spiritual authority into question. And yet, he also relates how the revelation instigates a realignment: "Now, say the children, now we know that the real Katchin will no longer ... dance in the pueblo. They will only come in an invisible way, they will dwell in a mystical way in the masks for the day of the dance" (p. 246).

For Žižek (2002), the revelation is a rite-of-passage that ritualizes the separation between what the children now know (that these rituals are purely symbolic) and what they believe (that they matter nonetheless). In essence, there is an acknowledgment of the limits of representation. In the same way that we all discover (at some point) that the tooth fairy is a representation, there is only so long the children can be fooled by the Kachin spirits. Thus, the community makes the revelation a formality. But as Žižek points out, the purpose of the rite of passage is not to dispel the children's belief but to shift its object. Rather than believe in the spirits themselves, they are asked to believe in the onto-symbolic order in which they appear. As Don Talayesva (1963) tells it, the revelation is a crisis of authority followed by a reaffirmation, a shift from supplication to responsibility. To understand the significance of this, we need to recognize the distinctive role of the masks. Often we think about masks as surfaces that hide. They are disguises that imply something more real or authentic behind them. Whether that be the horror of Ted Bundy (charming on the surface but a monster underneath) or the bravery of Don Diego de la Vega (a dandy on the surface but really the noble Zorro), the man behind the mask is of another order than the one we can see.

For Žižek (2002), however, the relation between the mask and the Real is differently prioritized. The Real is not the subject's inner truth but a negative force undermining the subject's capacity to be. The mask is the subject's positive persona; the representation of self-hood taken up through the onto-symbolic order; the image we need in order to imagine ourselves as self-standing coherent beings. In this rendering, the mask is what is significant. There is no deeper subject behind the mask – nothing more "real," except of course the reality of the Real itself. But the reality of the Real is not an anchor. It has no positive substance (it is a subtractive force) and only works to fragment and undo the subject's claim to itself. When understood in such terms, we can see why knowing that it is fathers and uncles behind the Kachin mask does not matter. What matters is how the mask gives form and appearance to the spirits themselves, allowing them to transpire as visible forces within the pueblo. To be sure, no one needs to believe that the spirits are real. This is not the role of the mask. It is not an event of transubstantiation. What matters is that they make the onto-symbolic world *appear*.

By providing a place for the spirits in the pueblo, the masks give the order a visible home, a material site in which to reside: a place where the order can be seen, and thus experienced as alive and present.

The second example comes from Benedict Anderson's (1983) discussion of the Tomb of the Unknown Soldier in *Imagined Communities*. While the bulk of Anderson's study on nationalist imaginations focuses on the advent of print capitalism, it is interesting that he opens the discussion with reference to something not only not textual but oddly devoid of determinative symbolic content. As he suggests, "No more arresting emblems of the modern culture of nationalism exist than cenotaphs and tombs of Unknown Soldiers. The public ceremonial reverence accorded these monuments [is] precisely *because* they are either deliberately empty or no one knows who lies inside them ... yet void as these tombs are of identifiable mortal remains or immortal souls, they are nonetheless saturated with ghostly *national* imaginings" (p. 19). My interest here centres precisely on this interplay between the emptiness of the monument and the saturated sentiments it engenders.

A similar relation is developed by Bataille (1985) in his analysis of the Egyptian obelisk in the Place de la Concorde in Paris. As Bataille suggests, the architect Jacques Ignace Hittorff was conscious of the Place's history as a site of both royal and revolutionary resistance and thus searched for a symbol that could not be appropriated by either side. The solution was the "apparently meaningless image" of the obelisk, a monument whose "calm grandeur and ... pacifying power" (p. 221) was able to weave itself into various dreams of nationhood, but in no distinct voice. Indeed, while geographers have a long history of looking for the meaning of monuments, symbolic content is often not the central purpose of public buildings. In Forster's (2002) discussion of the Neue Wache memorial in Berlin, for example, he illustrates how the central discussion of the memorial was about whether it should or should not signify. As the struggling republic debated how to memorialize the losses of the Great War, the debate centred on whether the shrine should focus on the living (and thus embed symbolisms of sacrifice, the nation, future generations, and the Fatherland) or be a site of mourning. Many commentators believed that the emphasis should be on mourning and, thus, that the memorial should not *mean* anything at all. As the liberal architectural commentator Siegfried Kracauer wrote in 1930, "What is required is not the zealous representation of content ... but the most extreme abstention toward it. A memorial for the fallen of the World War may ... not be permitted to be much more than an empty room" (quoted in Forster, 2002, p. 538).

The point I am making is not that monuments do not signify. They can and often do. But they are not *built* to signify. Rather, they are built to make the state present. Like the Kachin masks, the Neue Wache, the Tomb of the Unknown Soldier, and the Place de la Concorde make the state appear in the pueblo. They are material practices that allow the state to be experienced not as some symbolic idea (with its concomitant systems of meaning, purpose and allusion) but as a real and present force whose abiding concreteness imposes itself (obviously and forthrightly) as simply there. Such a conception of monuments is very different from cultural geography's traditional approach. Historically, monuments, memorials, and other built features of the landscape were understood in terms of their meaning - as symbolic objects that telegraphed various ideas about who we are through their symbolic forms. The point here is not that meaning does not matter or that subjects do not contest the landscape. The point is that these contestations are already invested in what the landscape makes present – that is, the state, culture, or community that the landscape marks. To focus on the meaning of landscape, therefore, is to misunderstand the nature of the landscape's power. The Cenotaph works because it creates a site where various ideas of the nation can gather. Its power comes from the diversity of meanings and attachments it collects rather than from the (ostensible) singularity of its symbolic form. Indeed, the more ideas about the state (its meaning, its purpose, its future, its legacy) that circulate around the Cenotaph – the more struggles and conflicts the Cenotaph instigates and creates space for – the more powerful the Cenotaph (as a marker of the nation) becomes. The Cenotaph (as a magnet for contestation, conversation, appropriation, and interpretation) makes the nation real – it makes it appear *as a thing*. This is geography's unique capacity: not to represent but to make present.[9]

To conclude this section, the discussion of masks and monuments has been about illustrating the fundamental role geography plays in constituting our existence as subjects. Claims are not just symbolic. They are material. They are built. In addition, it is through the practice of building (monuments, landscapes, plazas, etc.) that the onto-symbolic world is made present; even if that presence is vague and undefined, even if that presence transpires "in a mystical way." When I think of the rituals around British Remembrance Day, it does not matter that I personally do not participate in them. I do not wear a poppy and have only rarely remembered (or noticed) the moment of silence on 11 November. But a day or two beforehand, I usually recognize that something is happening. The sudden appearance of lapel pins, the veterans selling poppies on the street; these activities are material reminders that Remembrance

Day is here – that it is taking place – *with or without me*. This is the significance of geography. The material creation of symbols – of remembrance, peace, patriotism, sacrifice, anti-militarism, resistance – collectively work to create a place where the state can appear. The point of these geographies is not that they mean something. What matters is that they are present. Even when the nation-state is rejected, resisted, or damned through these sites, *the state remains visible* – existent because it is seen. This is geography's sacred work: making the onto-symbolic order real by making it appear.

Geography as Fabrication

It is hopefully clear at this point why geography is conceptualized as fundamental to the constitution of culture. Geography makes culture present. Rather than conceptualizing everyday cultural geography as something that reflects or expresses the internalized meanings of already acculturated subjects, the argument here is that geography matters because it allows culture to appear. It is through the production of material forms that culture (as a claim) transpires in our everyday world. Subjects do not have culture but claim it. And they claim it through building. Indeed, it is precisely because subjectivity is so fragile, because its claims are illusory and always already failed, that the material world is the go-to means by which subjects anchor their existence. Words fade and rituals are forgotten but the landscape is visible and durable. Subjects use the earth, rock, and mortar of our surroundings to write ourselves into the world. Is this not a ubiquitous human condition? The use of material resources to evidence that we indeed exist? As Zupančič (2017) suggests in her analysis of the performance of femininity, the subject's fear of losing her mask is not a fear of being found out – the fear of being revealed as not feminine (or not feminine enough). It is the fear of not having a mask at all: "The really troubling question here is: what if I'm not really anything, what if there is no 'me' in any of this? This ontological anxiety doesn't stop at 'Am I that name?,' rather it revolves around 'Do I exist at all?'" (p. 56). If the mask is what allows the subject to exist, then geography is that mask writ large in the world; not one mask among others but the fundamental mask, the mask of the earth itself, signposting how to be sovereign beings.

The final question of this chapter, then, is, How should we approach geography? What does it mean to analyse, critique, resist, or overcome the various social injustices perpetuated in and through everyday geography if it is all simply illusory; the manifestation of an onto-symbolic order that all subjects (in order to be subjects) are invested in? This is a complicated question that I will not answer conclusively. But I will however offer two signposts. The first is that it means approaching geography as a fabrication. I use this term consciously in order to capture both its senses. To say that geography is a fabrication is to argue that it is something manufactured.[10] This has been one of the central themes of the chapter, that geography is a unique modality of representation because it is built. It is geography's materiality (its manufactured nature) that makes it an effective technique for cementing claims. But the term fabrication has another sense. To fabricate is also to make up or to invent. This captures the illusory nature of geographical representation. Geographical fabrications narrate a story about who we are and how we exist. In calling this story a fabrication, I am not using the term pejoratively. These are necessary stories; stories subjects tell in order to claim themselves as sovereign beings. And while there is plenty to critique about the way some subjects tell these stories at the expense of others, one must understand the driving force behind such positive inscriptions. As struggles they are not only about power and domination. They are also (and more primordially) about being versus non-being. They are a struggle for existence.[11]

This leads to the second, and perhaps more important signpost. Approaching geography as a fabrication means not being seduced. While the geographical world we see may reflect the designs of hegemonic actors (and their various instruments of power), this is not to say that we should see those representations as signs of their power. Inscribing the world is a desperate act. It signals a hope – a prayer – that the sovereignty we desire can be actualized simply by writing it in stone. Is this not paganism at its purest? Building an image to stand in for the real? To be sure, representations have real effects. Fabrications injure, sideline, break, expel, exclude, and kill. The question is, To what extent do we endow those effects with the reality they claim? If we look back at the last century of cultural geographic analysis, is it not itself a symptom of geography's material seductiveness? From the advent of cultural geography, scholars have been analysing the landscape for evidence of an underlying *culture*. In essence, we have been analysing our own material inscriptions as evidence of our own existence. This is the seductive power of geography. By marking our existence in the landscape (and struggling over those markers), we essentially reinforce that there is something there, a real culture that needs to be defended, strengthened, resisted, or overcome. We as geographers have been party not to culture's unmasking but to its constitution. Or its constitution by way of its unmasking. Either way, we are part of the illusion.

In this sense, the politics I am advocating here (though in abbreviated form) is a politics of not being seduced; a political point developed by Lacan (2008) in his interpretation of the final scene of Sophocles's play *Antigone.* For Lacan, Creon invites the tragedies that befall him not because he interprets the law too harshly or is too inflexible in its application, but because he sees the law as a world apart. Rather than recognizing it as a representational order founded in response to the Real (and all the unconscious desires the Real obtains), he sees it as having its own logic, its own order, its own rationality and rationalization. He is seduced by the law's representational logic whereby it stands above – if not wholly outside – the alterity that gives it cause. Antigone, on the other hand, makes no excuse for her transgression. She does not deny the law's justice. She simply refuses to be seduced – to justify herself in the law's own language. My point, to be clear, is not to advocate for a politics of Antigone (which is a politics of self-destruction). It is more accurately to avoid the path of Creon; the path that is seduced by the law's own representational circle. For once we are in that circle, we can only reinforce its power, even when we are critiquing it. Not being seduced means recognizing fabrication for what it is: not a potent self-actualizing power but a feeble (though hopeful) claim.

Conclusions

The arguments presented in this chapter represent the final step in the overall thesis – that is, that culture is a claim and that geography is fundamental to its constitution. The argument has progressed through three key steps. Step one involved changing the question of culture from a question about how people exist to why they claim that mode of existence. Step two involved theorizing the subject through the concept of dwelling. The final step, which I draw to a close here, has been to illustrate how this conception of dwelling (and the subject who dwells) reorients traditional conceptions of culture and cultural geography. In the previous chapter, I argued that cultural ontologies need to be thought as forms of a response to the woundedness of being a living being. Culture, therefore, is how subjects chart paths through a mysterious world that diminishes, enfeebles, and undermines at every turn. In this chapter I have argued that these ontologies needs to be *built.* Culture transpires by being present. It is the visibility of culture that allows subjects to dream themselves as having some sovereignty – some claim – over their lives and their world. It is only when we understand the critical role geography plays in the constitution of culture that we can begin to see why I define culture as a dream of presence. Culture is a dream in the sense that it is an illusion of mastery, "this illusion of unity, in which a human being [is] always looking forward to self-mastery" (Lacan, 1953, p. 15; see also Chanter, 1998). But it is an illusion maintained through the work of visibility. It is geography that makes culture appear as a present and abiding phenomenon. It is geography that makes culture not present but a dream of presence – the dream of having a culture, an identity, an ontology, a world that is one's own.

The next and final chapter constitutes a conclusive statement on what I mean when I call culture a dream of presence. While the ideas should be relatively clear, the final chapter provides a succinct account of what the phrase means and why it is useful. It does not revise the arguments or provide a summary of key points. All the argumentative work is done. The final chapter is the shortest chapter and, frankly, has been the easiest to write. Freed from having to justify and tie together each and every position, it takes advantage of the opportunity to state simply and clearly the books central proposition: that culture is a dream of presence.

Chapter Nine

Dreams of Presence

Two Stories

Story 1: I was in a bad mood. I had been in Cairo since October, away from my partner for over six months and I had finally secured a decent flat where we could stay together.[1] She had been doing her own fieldwork in the United States and was supposed to be visiting next month. But I just got an email from her telling me she was delaying her trip. It was all very understandable but so was my annoyance. Plus it was hot. Full-on Cairo summer heat. And I was now late for an interview in Garden City. While the trip was relatively easy on public transport from my flat in Doqqi, I decided to catch a cab to make up time. I gave the driver the address but as I stared out the window, I noticed the driver was taking me in circles. "Do you know the address?" I asked in Arabic. "The address you gave me is wrong," he said, smiling, "but don't worry, I will get you there." Angrily, I told him the address was right and preceded to give him directions. His response was to pretend to not understand and turn down wrong streets on the pretence that he was just following my directions. Finally, I told him to stop, gave him 4 Egyptian pounds, and began walking to the interview.

At that point the cabbie came out of the car and told me I needed to give him 15 Egyptian pounds. And then … I totally lost it. I began yelling, calling him every bad word I knew in Arabic and he reciprocated, screaming he would kill me with his little finger. At this point, the merchants, doormen, and pedestrians of the neighbourhood began to gather around. Some began to rub my back, telling me it's okay, others were doing the same to the crazy cabbie. At some point in the scene, the driver said he was not leaving until he received another 1 Egyptian pound and one of the men looked at me and cocked his head, "give him the pound" it suggested, "this is ridiculous." I left the scene flustered and angry. But most of all, I was embarrassed. The men who had gathered around were beseeching us to be respectable. "Calmness," they kept saying, "civility." In a reversal of the classic colonial discourse, I was the childish, uncivilized, brute whose mad antics were not only embarrassing (for myself and them) but were disrupting the ordered civility of public life. My cab driver was acting crazy. But I was no better for responding with an equal amount of intensity. This is not the way civilized people behave, the crowd told us, control yourselves and be civil.

Story 2: Dotted along the Pyramids Road are a number of small unadorned wooden signs: "Cleanliness is a sign of civilization,"[2] one says, and "Listening to the radio loudly is a sign of being uncivilized" says the other. There was another near my flat: "Pay attention to the traffic officer, he is demonstrating progress and civilization," and another in Saqqara off the Mansouraya Canal: "Egypt is the land of civilization." During the course of my year in Egypt, I heard many informants discuss cultural heritage in terms of "culture and civilization," but it was not until I noticed the signs that I began seeing how the terminology was diffused through a wider social and political discourse. In the last week of my fieldwork in September, I decided I wanted to take some photos of the signs and began travelling around the city capturing images. On my last stop, there was road work being done and the sign had been taken down and set sideways on the pavement. It was not ideal but I decided to take the shot anyway.

As I crouched down composing the photo, one of the people walking past me stopped and blocked the shot. I looked up and there was a man asking me what I was doing. I told him I was doing research on the concept of civilization. The man became agitated: "They took it down so they could make renovations on the street," he said. "They will see your picture with this sign on the ground and they will think bad things about Egypt – that Egypt doesn't care." I tried to assure the man that it was not my intention to denigrate Egypt but to show how Egypt cared about civilization. But the man was not satisfied: "If you were renovating your house, how would you like it if I came and took a picture when it was halfway done?" I again attempted to reassure him that I was not going to "say bad things about Egypt," and that if I had time I would take another picture when the sign had been put in its proper place. Again the man was unmoved: "It depends what kind of person you are. If you want to take the picture, go ahead."

I tell these stories because they tell us something about the way the notion of civility circulates in Egypt. The Arabic root of the word for civility, civilization, and civil is *hdr*, most often rendered in the word *hadaara,* which means civilization as well as culture. The pyramids, for example, are often narrated as a sign of Egypt's *hadaara,* its ancient civilization, its sophisticated culture, and its long-standing rootedness in an ancient lineage. But what is interesting about both stories is that civility is mobilized as a narrative just as it appears to be on the cusp of being lost. In the first story, it emerges in response to the threat of civil disorder and violence. Indeed, the scene I was making with the cabbie was not simply unsophisticated, base, and unseemly, it was irruptive and unpredictable. Here was a crazy *chowaga* fighting over an Egyptian pound (back then about 20 cents). What else might he do? In the second story, I was introducing a photograph into a world whose audience could not be controlled. I did not propose or want to embarrass Egypt, but the man had a point in the sense that there was only so much I could do about this. Regardless of my intention, the photograph *could* "say bad things about Egypt," my assurances to the contrary being empty and conceited. In this sense, his comment that it depends on what kind of person I am rings true. Both stories depended on what kind of person I was, a question I could not (and cannot) readily answer. As many Egyptians know (and as we privileged few often forget), we are at the world's mercy, vulnerable to its torsions and convulsions, held out to its violences and surfaces of exposure. *Hadaara* is one of the ways Egyptians respond; how they declare a self, vis-à-vis the crazy in the street or a photo in a country far away.

It is only when we see culture operate in these terms that we understand why hadaara cannot be understood as a discourse that inscribes an interior culture. While it is a norm, it does not work by disciplining the subject's body and mind. On the contrary, what *hadaara* signals is a path: an inherited symbolic form that Egyptians can use to claim themselves. And it is a path that is built. *Hadaara* appears everywhere in Egypt. This can be seen most obviously in the monuments that the Egyptian government spends enormous amounts of money managing, repairing, displaying, storing, and cataloguing. Antiquity in Egypt does not reside only in the outskirts of the city, in national museums, or in faraway cities like Aswan. Egypt's ancient, Byzantine, Islamic, Turkish, and colonial inheritance is an indelible mark on the state's material form. From the neighbourhoods of old and Islamic Cairo, to the open-aired markets, to the cemeteries that surround the city, to the modern office towers and hotels in the Central Business District, Egypt's *hadaara* is preserved, referenced, symbolized, and alluded to everywhere. Civility is not a matter of atmosphere. It is a matter of infrastructure. As a material marker of identity, it is written into the city's physical fabric.

The promise of this book – a promise perhaps I am only making good on now – is that it will offer a geographic theory of culture. While I believe that the basic tenants of this argument have been developed throughout, the aim of this final section is to spell out precisely what makes this theory distinctive and how it can be used to think about culture and identity in novel terms. I have organized the chapter around four central propositions, each of which compares the concept of culture I have developed against previous formulations of the term, particularly in geography. Thus, culture is (1) something desired (rather than imposed), (2) something built (rather than expressed), (3) a modality of faith (rather than power), and (4) a politics of tolerance (rather than a politics of representation). The chapter then concludes with (what I hope is) a usable and concise definition of culture – that is, culture is a dream of presence. To be clear, this chapter is summative in the sense that it distils the book's argument into a set of key points, most of which have been introduced already. The one caveat is perhaps the last proposition, which provides some all-too brief signposts towards the question of power and politics. While I recognize that these issues are not being covered adequately, I hope this brief discussion – as well as my previous writings on these issues (see Joronen & Rose, 2021; Rose, 2014, 2019, 2021b; Rose et al., 2021) – give some indication of future directions. My aim in this work, as suggested in the introduction, is to provide a working theory of culture, the potential implications and applications of which await further development.

Culture Is a Desire

For most of the concept's history, culture has been understood as an interiority. By this I mean, it has been thought of as something that resides within the subject, an internalized architecture of thoughts, feelings, ideas, values, and meanings that subjects carry around inside them. A name we could articulate wherever we were because it was always there; something we could whisper to ourselves in the dark, even if no one arrived to ask. Given this presumption, the task of cultural theory was to contemplate how this architecture was shaped, structured, and organized by external forces: What was the mechanism by which certain ideas, orders, and norms were imposed on the subject's consciousness to create this internal voice and its distinctive ways of knowing? For Bourdieu (1990), the mechanism is corporeal practice; for Geertz (1973), it is performative ritual; and for cultural geographers, it is the landscape itself, seducing us with its naturalness and beauty. In all of these examples, the engine of culture is society and the target is the person inside our heads; our worldview, our minds, our taken-for-granted presumptions; an internalized consciousness silently (and perhaps insidiously) shaping our behaviour and keeping us (more or less) compliant.

In the framework I have developed, culture is definitively not imposed. It is not something levied or stamped and there is no external engine imprinting our bodies and minds. On the contrary, culture is something sought, a refuge where subjects seek the semblance of sovereignty. The argument is predicated upon two key points. The first is that there are two separate questions within the question of culture: there is a question about difference and there is a question about claiming. The question of difference is unanswerable. While difference solicits endless interest, it provides no answers. Difference is indeterminable, unnameable, mysterious.[3] This leads to the second question of culture, which is a question about why people claim their difference. If difference is elusive and inscrutable; if it is something given rather than taken; if it is something that denies understanding (both our own understanding and the understanding of others), then why do we claim this difference *as if* it were ours? The central question of culture, therefore, is not about difference itself but about this claim to difference. Why do subjects claim the identities they have been bequeathed – given from a time out of time – as their own?

The second key point answers this question by developing a novel conception of subjectivity. Drawing upon Heidegger and Levinas, the book argues that subjects are never self-standing sovereign beings but beings who are beset, wholly and unrelentingly, by forces radically outside them. For both thinkers, subjectivity is never won. It is only pursued. The implications of this framing are twofold. At one level, it undermines conceptions of culture that begin with a sovereign subject who is secondarily shaped and moulded by society. From Rousseau's *tabula rasa* to Butler's notion of performativity, this idea of a pre-existing will or consciousness which precedes culture (and resists its various modes of inscription) has historically been the bedrock for thinking culture and society. This project, however, understands subjectivity as something pursued (something desired) rather than something given. Such an approach shifts the whole direction of cultural theory. Human beings are not primordially free and secondarily inscribed but are always and unrelentingly beset. Sovereignty, therefore, is an elusive desire and culture is a refuge where this desire is (at least imaginatively) claimed. This brings us to the second proposition.

Culture Is Built

Material culture has always been an important part of both anthropology and geography, and historically it has been an area of some traffic between the disciplines (Bender & Aitken, 1998; D. Miller, 1994, 2005; Tilley, 2004). While the emphasis of the former tends to be on commodities, tools, and sacred spaces, the emphasis of the latter tends to be on parks, landscapes, and urban planning. In both frameworks, the theoretical question concerns the dialectical relationship between the normative world outside and the individual subjects who absorbs that world. It is a framework that (once again) presumes the material world to be an expression of something already inside us, a culture already there working to structure and organize the everyday material world in its image (Rose, 2002).[4] In arguing that culture is something built, my aim is to reverse this relation. Rather than presuming culture to be an interiority that is secondarily expressed, I have argued that culture is wholly exterior, built in order to materialize the claim that we are sovereign beings. The argument is based upon a particular rendering of representation.

Traditionally, representation is thought to express not simply an underlying cultural identity but also a particular interpretation of that identity. Thus, the landscape does not represent culture but is a site where different interpretations of identity struggle over the right (and capacity) to materially represent that culture. This means that the emphasis of the analysis is on interpreting what a particular representation of identity *means*. In the framework presented here, the significance of representation is not about what a particular representation means but what it *marks*. What matters is that culture appears. How it appears (while potentially important) is secondary. Appearing is primary since it is only by being materially present that culture transpires at all. This is also why geography is fundamental. It is through the material construction of our physical world – by literally writing the earth – that culture comes to be present as an abiding phenomenon. In this rendering, geography can be understood as a kind of master play. While representation has no ability to actualize a subject's sovereignty, it does have the ability to represent it. By building the material world, subjects generate, sustain, and invest in an onto-symbolic order that reflects back to them a mastery they never properly have. This leads to the third proposition.

Culture Is a Modality of Faith

Much of late twentieth-century cultural geography was founded on the presumption that knowledge was a mode of power. Thus, representation worked because it had the capacity to shape our understandings of the world (our knowledge), which in turn allowed embedded power relations into the unquestioned order of things. In response, I have argued that representation does not shape knowledge but rather shapes belief. As Žižek (2002, 2006, 2008a) argues, subjects are very aware of the representational logics of their world. They know full well, for example, that money is a symbol and that commodities hide their true relations of production. They are not "mystified" by capitalism or fooled by its fetishizing images. Yet this does not mean that subjects do not act in concert with these representational logics. As Žižek (2008a) suggests, we are "fetishists in practice, not in theory" (p. 28). To understand this, we need to recognize that subjects are fundamentally invested in the onto-symbolic order. Representation, in other words, is not something subjects seek to escape but is something they are actively involved in composing. This is why Don Talayesva's (1963) discovery that it was his father and uncle behind the Kachin masks is not a Wizard of Oz moment. It is not as if the whole onto-symbolic order crumbles. On the contrary, what the revelation reveals is that the order is now his responsibility to sustain. This is the real reason why no one reveals to the emperor that he has no clothes. The issue is not simply the obsequiousness of his courtiers or the fear of his citizens but the fear of speaking a secret that everybody already knows – that the onto-symbolic order is fragile (Žižek, 2006). To call out the king's nakedness is to call out the illusory ontological order *as illusory*; a truth only the child dares to speak, the consequences of which are never explained.

The point here is to make clear the precarious and paradoxical manner in which the onto-symbolic order operates. On the one hand, the onto-symbolic order is necessary. It is how subjects come to claim themselves vis-à-vis the existential precarity of the Real. And yet, subjects are also (often painfully) aware that these representational acts have no hold on the world they claim. Subjects know the precarity of their claims (even if they are not open about it). We know that other people need, that illness inflicts, that the future haunts, and that hunger is never far away. To say that these truths are not part of our everyday repertoire of knowledge would be to ignore the full-array of clichéd aphorisms we repeatedly tell each other, for example, "life isn't fair," "shit happens," "it is what it is." Thus, even as building works to hide and mitigate the power of the Real (and in the process claim some sovereignty over our subjectivity), this does not mean we are unaware of its limits. This is what makes the onto-symbolic order not an order of power but an order of faith. It is something we are invested in despite its precarity. Is this to say that it has no power? No. As suggested in the previous chapter, the onto-symbolic order is both an order of identity and an order of calculation: a set of institutionally derived protections that are unevenly distributed. But this is precisely what makes the politics of representation a far trickier proposition than traditional representational analysis allows. This leads to the final proposition.

Culture Is a Politics of Living Together

Once we understand the onto-symbolic order as an order of faith, a very different understanding of contemporary political relations potentially begins to take shape. For many years, power has been conceptualized as the driving force behind all social relations. Thus, social injustice and exclusion are understood to be the result of the powerful using their advantage to organize the social order according to their will. To be clear, I do not deny that the onto-symbolic order is a system of power. As a representational economy, it works to secure access to resources and privileges to some while leaving others precarious and vulnerable. But this is not to say that power relations are the *source* of that vulnerability. As I (and Mikko Joronen) have argued previously, "vulnerability emerges not from bodies, affects or distributed agencies but from a situation that sits outside such relations and their terms; a situation no power can ever eliminate or incorporate; namely, from the woundedness of being a living being" (Joronen and Rose, 2021, p. 1410). Once we understand the woundedness of being a living being as the source of all social relations (including relations of power), then we also see how all social relations are a response to this condition. And we see that every will to power is an attempt to mitigate the insecurity it engenders. This means that power relations cannot be thought of as the source of social injustice. On the contrary, woundedness is the *a priori* injustice. And it is precisely because of this *a priori* injustice that justice is distributed unevenly.

Such a perspective creates powerful consequences for understanding the power of representation and the problem of living with others. In terms of the former, the power of representation becomes a far more ambivalent affair. When the engine of representation is power, it makes sense to undermine the work of the onto-symbolic order because it is designed to sustain social hierarchies. Identity, in this framework, is simply a proxy for power relations. Thus, challenging symbolic constructions that uphold privilege and injustice is a legitimate political act. But when the engine of representation is faith, the political calculation changes. Having faith in the onto-symbolic order means having faith in that which sustains one's capacity to be a subject. And while this order is no doubt open to being challenged – every claim to identity is (after all) only ever a claim – what is being confronted is not just power but the onto-symbolic order more broadly, that is, the very material by which subjects define their subjectivity. In other words, while it is easy to undermine a representational claim if it is understood as a vehicle for social domination, it is less easy to do so if it is understood as the very means by which subjects become subjects, particularly when that system is already precarious and under siege.

In terms of the problem of living with others, conceiving power as a response (and identity as something precarious) illuminates the sheer difficulty of living in community. While I used Levinas (1969) earlier to illustrate how the arrival of the other person is the advent of self-consciousness (an event that Levinas understands as the origin of all ethics), it is important to recognize that this encounter is deeply unsettling. Living with others means being among those that are not simply racially, ethnically, or habitually different, but more importantly, those that are psychologically, dispositionally, or attitudinally "other." Different in their needs, desires, and demands. And while we are no doubt in relation with these others, we need to remember that we are also not in relation with them. There is always a dimension of the other that recedes from our grasp. Thus, for all the emphasis on illuminating our connectedness to other beings (e.g., in Haraway, 2008; Latour, 2004; Stengers, 2010; Tsing, 2015), there seems to me to be an unwillingness to acknowledge that there are also infinite distances; gaps that not only keep "me" separate from "you" but that keep me from fully incorporating you into my being, my relations, my networks, my community, and my politics. On the one hand, it is this distance that allows otherness to abide as otherness, that is, that keeps others from being fully incorporated and colonized. But it also allows otherness to thrive in its other unincorporated distance, sometimes taking on monstrous forms (Vollebergh, 2016).

Thus, rather than attempting to undermine representational economies or incorporate others through relation, I see this project as pointing towards what I would term a politics of *tolerance*. One has to be careful when using this term, as *tolerance* has undergone extensive critique over the last decade (Brown, 2006; Brown et al., 2014; Žižek, 2008a). As Wendy Brown (2006) argues, tolerance is a term society uses "when a group that poses a challenge … must be incorporated but also must be sustained as a difference" (p. 47). While in Brown's framing, this desire to incorporate but also sustain difference is a problem, I see it as a necessity. Political incorporation must recognize an unassimilatable difference. In a similar manner to Mouffe (2013), I argue that there is no getting around the problem of difference. To incorporate difference is to colonize it. And to reject it is an act of violence. Tolerance, therefore, stands as a reasonable political position. It means remembering that every claim to identity is a claim to sovereignty and, simultaneously, it is a claim against others. Trying to balance such claims in a way that builds justice is a perpetual struggle. A struggle against the claims of others but also a struggle against the *a priori* condition that makes such claims necessary; indeed, that makes all claims necessary. Tolerance, therefore, invites a certain compassion for claims, even when they are monstrous. It does not mean accepting the claims others claim, but understanding the dark background from which they arise, even as we may wage war to defeat them. Tolerance is thus a call for empathy without sympathy. It rejects the easy sentimentality of multiculturalism (a project still predicated on a dream of human sameness) and accepts that war is perpetual and obligatory, all the while reaching for less violent means. As Levinas (1969) says, we cannot kill our way out of the problem of other people. Otherness will always be there. Tolerance is a politics that accepts this condition, as well as the burden of conflict it necessarily invites.

Culture Is a Dream of Presence

Taken together (and by way of conclusion), these four propositions provide us with the following working definition of culture: *culture is a dream of presence*. Throughout the text I have used different terms to discuss the frailty of the onto-symbolic order. I have called it illusory, mysterious, and fabricated. The reason I am only now switching to the terminology of dreaming is because I think it is only at this point that the implications of this representational conception of culture can be fully appreciated. On the one hand, the terminology of dreaming is consistent with the vocabulary I have relied upon thus far. To call something a dream is to characterize it as ethereal, elusive, chimerical, and perhaps even delusional. However, the term *dream* also carries with it another dimension. To dream is also to hope. It is to pursue something despite the odds. This is the dream of Willy Lowman and of Jay Gatsby, a pipedream that summons from a distance but whose collection is perpetually deferred. Thus, to say that culture is a dream of presence is to say that it is an illusion but also a necessary illusion. (Is it necessary? It is necessary.) It is an illusion that beckons from afar and instigates subjects *to be subjects*, that is, to be beings that make claims about *their* life, *their* bodies, *their* self, vis-à-vis the subtractive force of alterity.

Thus, while culture is an inheritance – a gift from another time – what this inheritance provides has no automatic hold on the world or on the mind. All it can do is summon. We can never reach culture, we can only move towards it. We can never have culture, we can only claim it. This is why culture requires some site – a material screen – upon which its dream of propriety can be projected. Given that culture can never be interior (it can never be given a stable home within the subject) it requires fabrication (totem, theatre, or landscape) where our dreams of being an acculturated subject can be displayed. Hence, culture is fundamentally geographical. Geography is the means by which subjectivity comes to appear as substantive and real. "There is no ipseity without some prostheticity," Derrida (2009, p. 88) writes. There can be no subject (no "I am") until there is some predicate upon which my existence as a self-standing being can be transposed. Even when we recognize such transposition as fabrication, it is nonetheless necessary. Culture (as claiming) is a different kind of earth writing. One perhaps that geographers have always been attuned to but never took to be fundamental. Culture is not just a dream. It is a dream of *presence*; a dream that relies upon being written into the world. Geography is necessary. *It is necessary*.

Notes

1. Culture War, Culture Loss

1 See for example discussions of victim culture (Brooks, 2015; Friedersdorf, 2015a, 2015b; Lukianoff & Haidt, 2018; Manning & Campbell, 2014; Saul, 2017; Sommers, 2000), cultural appropriation (Avins, 2015; Friedersdorf, 2015a; Frum, 2018; Malik, 2017; Pham, 2014; Sehgal, 2015; Will, 2017), cultural politics (Hu, 2020; Jackson, 2020; Traldi, 2020), cultural anxiety (Douthat, 2010; Green, 2017), corporate culture (Blum, 2020; Campbell & Strachan, 2018; Greg, 2015; Peck, 2018; Stalans & Finn, 2018; Washburn, 2020; Williamson, 2020), conspiracy culture (Goldberg, 2001; Lafrance, 2017; Uscinski & Parent, 2014), and of course cancel culture (Donato, 2020; Douthat, 2020; Friedersdorf, 2017; Hobbes, 2020; Lewis, 2020; McWhorter, 2020; Ponnuru, 2020; Ross, 2019; Waldman, 2019; Weiner, 2020; Yar & Bromwich, 2019).

2 For more on Christian nationalism, see Baker et al. (2020); Whitehead et al. (2018); Stroope et al. (2021); Davis and Perry (2020); and Goldberg (2016).

3 This is not to deny that culture involves various social contestations and struggles for dominance. Yet this framework presents these as secondary effects of a more primary existential situation.

4 In many ways this work sits alongside a number of recent writings responding to what Philo (2021b) terms *the new positivism*: "a resolutely affirmative, positive tonality of research that balks at expressions of negativity" (p. 5; see also Philo, 2021a). While much of this work takes its inspiration from the Levinasian inflected writings of Harrison (2007a, 2007b, 2008, 2009, 2015) – see, for example, Bissell (2022, 2023); Bissell et al. (2021); Dubow (2011, 2021); Rose (2011b, 2014, 2021b); Wylie (2009, 2012, 2016a) – other emerging critiques of affirmationism can be seen in the work of Pugh (2021, 2023); Zhang (2020, 2021, 2023); Dekyser (2021, 2022); Joronen (2021a, 2021b, 2021c); Joronen and Rose (2021); as well as Ash

and Simpson (2016); Chandler (2013); Secor and Kingsbury (2021); and Strohmayer (2021).

5 This point is made particularly brutally by Žižek (2002) who characterizes Cultural Studies as an ani-intellectual avoidance strategy predicated upon unacknowledged "ontological and epistemological presuppositions on the nature of human knowledge and reality: usually a proto-Nietzschean notion that knowledge is not only embedded in, but also generated by a complex set of discursive strategies of power (re)production, etc." (p. 298).

Section 1. Culture

1 Livingstone (1993) notes that Hartshorne acknowledges regions to be "mental constructions" artfully construed by the geographer. This is an idea that resonates in the presidential address of John Fraser (Hart, 1982).

2 It is important to distinguish here between explanation and description. My point is about the former – difference itself cannot be explained. We cannot know what makes a person, community or nation different. But this is not to deny the importance of ethnographic description and trying to understand (partially) the perspective of the other. Here I think Viveiros-de-Castro's (2004b, 2014) writings on controlled equivocation are essential reading.

2. Anthropology and the Naming of Difference

1 The terminology of "native" is controversial as it is differentially refused and/or appropriated by Indigenous communities and individuals. I use it because it is common parlance within much of the anthropological literature I am citing.

2 Viveiros de Castro (2014) criticizes what he terms the *complacent paternalism* of this position, which "simply transfigures the so-called others into fictions of the Western imagination of which they lack a speaking part" (p. 40).

3 The circularity is similar to the one discussed by Foucault (2003) in his conception of governance. For Foucault, states create various technics of power to solve problems of their own creation. Thus, the technics of biopower are devised as a response to the ostensible problem of "population" (Foucault, 2007; see also Rose, 2014). It is the invention of population as a problem that both justifies biopower and (simultaneously) reinforces "population" as a problem in need of a state response. In this sense, it could be argued that anthropology "problematized" culture. It made culture into a problem and then devised various techniques (e.g., ethnography) for solving it – thus reinforcing culture's problematic nature. I discuss this circle further in chapter 7.

4 The anthropologist Roy Wagner (1981) makes a similar argument but comes at it from a very different position. For Wagner, culture is indeed an invention that subjects use to explain the difference of others; a difference anthropologists are forced to reckon with through the ethnographic encounter. By calling the culture concept an invention, Wagner draws attention to the way anthropologist and native alike invent ways of categorizing themselves and the other, creating relations of sameness and difference. While anthropologists do not see themselves as having culture, through the ethnographic encounter they come to see their own differences. But these differences are semiotic improvisations, akin to Derrida's concept of *differance*. For more on Wagner, see Holbraad and Pedersen (2017) and Dulley (2019).

5 It should be noted that much of the work on Indigenous ontologies in geography could also be characterized as falling within this second camp. Certainly, in the work of what might be characterized as the "Australian School" of Indigenous studies, there is an interest in how Indigenous ontologies (conceived as belonging to Indigenous people) can help displace, correct, or otherwise inform Western or colonial perspectives.

6 I am compressing quite a bit here. There are differences between Strathern (who leans on Haraway) and Viveiros de Castro (who is more Deleuzian) and others such as Pederson and Blaser, who draw more obviously from Latour.

7 While I recognize that ethnography is a distinctive method for encountering otherness (bearing the imprint of a distinctive epistemological outlook and method), the practice of naming one's own difference in relation to others is by no means extraordinary.

8 One can see this tension playing out in a well-known debate between Graeber (2015) and Viveiros de Castro (2015). While the former argues that "there is never any sense that people existing inside other Ontologies have any trouble understanding each other, let alone the world around them" (p. 22), the latter responds that ontologies are not "rigid, impenetrable solids that are solipsistically 'withdrawn' within their own incommensurability" (p. 10).

9 I would suggest that Holbraad and Pedersen's (2017) unwillingness to engage with the problem of "who" is related to their dedicated avoidance of attaching themselves to a particular metaphysical camp. While I am sympathetic to the recognition that every metaphysical story ultimately fails, I would add that this fact does not let us off the hook – that is, one cannot simply circumvent or disown the work of metaphysics. As Michael Roth (1996) once pithily put it, "if you leave metaphysics to take care of itself, it will eventually take care of you" (p. 12). There is always

a metaphysics present in what we write. Thus better to acknowledge our position than to deny that we have one.

10 When I first began presenting these ideas many years ago at a seminar at the University of Wisconsin, an eminent professor invited me for a chat and posed an important question which – at that point – remained unanswered (or certainly unanswered clearly): "Who dreams?" I am essentially posing the same question.

11 Graeber (2015) actually makes a similar point in his debate with Viveiros de Castro: "It seems to me that taking one's interlocuter seriously means, not just agreeing with everything they say … but starting from the recognition that neither party to the conversation will ever completely understand the world around them, or for that matter, each other" (p. 27).

12 Such an argument no doubt could be seen as politically disempowering as it robs Indigenous groups (or indeed any group) of their difference – or as Schor (1989) puts it, a right to their difference. To be clear, I am not arguing that others are not different. But I am arguing that those differences are unknowable and thus any mobilization of them would be a form of strategic essentialism (Spivak, 1996), which of course is a valid politics. I talk more about this (though in brief terms) in chapter 9.

3. Geography's Cultural Legacy

1 Jovan Cvijić, *La peninsule Balkanique: Geographie humaine* (Paris: Librairie Armand Colin, 1918), quoted in R.D. Campbell, "Personality as an Element of Regional Geography" (1968, p. 749).

2 When Ratzel talks of *kultur* he does so in the sense of civilizational status or position rather than in the way we think of it now, with its association with norms, habits, traditions, ritual, and so on. Thus, in his work on the Mediterranean, he explores the traces of old (read noble) Roman *kultur* through the territories of the Levant. For this reason, we will mostly avoid Ratzel's commentaries on *kultur* in favour of his discussion of *volk*.

3 It is surprising that the rise of habitual, practical, and relational theories in geography have not referenced the geographical traditions from which such ideas arose, particularly as they are often celebrated as the crucible of relational ontologies. As Malpas (2008) and others have illustrated, many of the theorists celebrated in contemporary geographical work were themselves powerfully influenced by the anthropogeographical tradition. Heidegger was a keen reader of Vidal (Malpas, 2008) and a number of authors have drawn parallels between Deleuze and the French geographical tradition (Lindaman, 2017; Merriman, 2012). In addition, while Ratzel is not cited as a direct influence, Verne (2017) makes the connection between his work and poststructural process theory clear:

"When we now draw on Deleuze and Guattari, and Latour and DaLanda to inspire our apparently new and highly innovative reflections about the mobile, procedural and relational nature of spaces – all thinkers directly referring to Spinoza and Leibniz among others – hardly do we realize that it was these very same scholars that were an important influence for Ratzel" (p. 90; see also Malpas, 2012).

4 I am wary of making this point to strongly as I think there is a case to be made that Sauer (like Ratzel and Vidal) anchored his conception of culture in a notion of livelihood. As A.H. Clark (1954) suggests, one dimension of Sauer's work that is often overlooked is his enduring interest in economy: "The largest single theme … clear in the writings of … the Berkeley group … has been the emphasis on … man's use, alteration, and rearrangement of his only potentially permanent resources: water, soil, vegetation and animal life" (p. 89). This emphasis can be particularly seen in Sauer's (1941) work on Mexico where the diffusion of productive technologies is often a key criteria in his determination of cultural hearths. In addition, if this characterization is perhaps a stretch too far for Sauer, it could certainly hold for Sauer's students, particularly those associated with cultural ecology. In Denevan (2001), Butzer (1993), Zimmer (1996), and others, landscapes reflect not simply cultural patterns but cultural patterns of material production (see Mathewson, 2011). As Robbins (2004) suggests, the aim of cultural ecology is to understand how groups approach and solve environmental problems (related to systems of production) and how those approaches illustrate themselves in the landscape. In this sense, cultural ecology cracks the black-box of culture, even as it leaves the "culture as pattern" framing largely intact.

5 For regional geographers, the mechanism for this mindset were the unique characteristics of the *genre-de-vie*, a set of human-environmental interactions that gave people in the community a particular "view" or "way of seeing" the world which then impacted the way they appropriated new information.

6 To be clear, I mean a high point in cultural geographic work that *remains interested in the question of culture*. Non-representational theory could also be called a high point of cultural geographic work. But as a mode of doing cultural geography, it is not particularly interested in culture nor (obviously) in representational analysis.

7 I am unsure whether my contention that Foucault had an outsized influence on cultural geography needs to be defended. But if so, I can think of (at least) five entry points where we see Foucault having a significant influence on the subfield. The first is through the historical work of Driver (1985, 1997), Philo (1992), and others (e.g., Ogborn, 1992, 1995), who, as Hannah (2016) notes, were working with Derek Gregory at the University

of Cambridge on issues of governance and institutionalization (see also Hannah, 1993; Matless, 1992a). A second strand is identified by Barnett (2001), whereby Foucault enters the lexicon of cultural geography via the more generic influence of Cultural Studies, which becomes increasingly interested in issues of bodies, institutions, and cultural artefacts. A third entry point is the "imaginative geographies" literature pioneered by Gregory (1994) but furthered by Duncan & Gregory (1999), Driver (1992), Driver & Gilbert (1999), and others (e.g., Gruffudd, 1995; Jackson, 1998; Jazeel, 2014; Matless, 1994, 1998; Pratt, 1992) who were influenced by Said's (1979) unique interpretation of Foucault in *Orientalism*. Fourth, is an interest in Foucault's genealogical method that influenced the historical geographical work of Matless (1998) and Hannah (2000). Finally, there is Foucault's broader interest in space and "locations of power" (see Foucault, 1986), which influenced many areas of geographic work (e.g., see Soja, 1988, 1996) but had significant impact on cultural geography.

8 These terms are borrowed from Bourdieu (1990), Foucault (1995), and Gramsci (1995), respectively.

9 As suggested in chapter 1, Mitchell (1995) makes a similar critique, though I see this problem as having less to do with culture's utility as a concept and more to do with its unique history of conceptualization in the discipline.

5. Heidegger's Dwelling

1 While the mercurial (even snide) "so it seems" would seem to militate against this reading, one needs to recognisz the broader context. I would suggest that this is a dig at the "they" – that is, the masses who perpetually misunderstand the fundamental problem. Heidegger's point is not that dwelling is not attained. His point is that it is attained by dwelling rather than building.

2 This is not precisely accurate. Dasein is the name Heidegger gives to that being who is concerned with their being and the presence and absence of equipment. In this sense, it could be argued that Dr. Cornelius (the scientist character from the original *Planet of the Apes* film) is Dasein and it could also be argued that a human with severe mental disability is not Dasein. As we will see, this terminology changes in his later work where the term *mortal* takes on greater significance.

3 Heidegger makes clear that his main interest in *Being and Time* is not an analysis of Dasein per se but an analysis of Being for which Dasein is a means to address (see Heidegger, 1996, esp. division 2, chap. 3).

4 Heidegger (1978b) later calls this harkening "the pull" of being, that which "calls on us [and] demands for itself that it be tended, cared for husbanded in its own essential being" (p. 367).

5 It would be more accurate to say the *Augenblick* reveals Dasein's temporality, rather than its historicity. I have chosen the latter in order to emphasise the historicity of the world, i.e., its emergence through a series of temporal events that Dasein has been bequeathed. But the emphasis in *Being and Time* is on Dasein's reckoning with finitude and mortality; a dimension I am eliding somewhat in order to keep the line of argument clear and avoid introducing too many of Heidegger's concepts.
6 It is this capacity to see the world as temporal that distinguishes the human from the animal and, thus, what makes Dasein of a different ontology (of a different mode of being). As Heidegger (1995) discusses in his 1929–30 lecture course, animal absorption in an environment is not illuminated by the *Augenblick*. Thus, there is no moment where animals can behold the historical deliverance of their world.
7 There is some debate about what the *Augenblick* actually changes, for example, see Joronen (2021a).
8 Though also political. see *Heidegger and Language* (J. Powell, 2013a).
9 There is a justification for using Being in its capitalised form here since we are now talking about Being in its "Big Being" sense. However, I feel that would raise questions about the difference between "being" and "Being" which I am not directly addressing in this chapter and would take me far away from its immediate concerns. Thus, it would create more confusion than clarity for readers not familiar with Heidegger's broader oeuvre.
10 Heidegger's engagement with the concept of *physis* continues throughout his middle and late writings on technology (1977a), poetry (1971b, 2000), art (1971c), thinking (1977b, 1978b), and language (1971b).
11 Many commentators hear in these words the rumble of Nazi tanks (see Wolin, 1992, 1993).
12 Malpas (2006) also sees the gods-mortal axis as temporal and the earth-sky axis as spatial.

6. Dwelling as Both Open and Closed

1 It should be acknowledged that for both Heidegger and Levinas the dwelling mediates between the centripetal forces of subjectivity versus the centrifugal forces of background relations – understood in Heidegger as "the world" and in Levinas as "the other." While their metaphysics are different, the dwelling concept serves both frameworks in a similar fashion.
2 There is a lot of obvious overlap here with Lacan, some of which I will discuss further in chapter 8.
3 The concept of nowhere recurs in Levinas's thought in different guises. It is best to think of it as a dimension that is non-present, that is, it has no

spatial or temporal place in the world but exists in a non-transposable dimension.

4 It should be noted that this account of subjectivity has a certain mythic quality to it. Subjectivity, in this rendering, is something that can only be captured retrospectively. In this sense, it reads a bit like an origin story – less Homeric epic and more Levi-Straussian cosmology.

5 This dimension of exteriority is a key reference point for Levinas, particularly in *Time and the Other* (1987). Here, time is of a different dimension. It is not a river or flow of events because these are merely reflections on and projections of the present. For Levinas, time is of a different order and dimension. It arrives from somewhere else – a nowhere that is not of the present and can never be of the present.

6 One can see some obvious parallels here to Laplanche and his concept of ego formation discussed in chapter 4.

7 Levinas does not refute Heidegger's argument that subject's care for themselves and their world. What he refutes is the primordiality of that position. For Levinas, the ego's self-identification as an 'I' is precipitated by the other person rather than by its sense of being unhomely.

8 While this chapter maintains a certain fidelity to Levinas' argument, the distinction he draws between humans and animals (see for example, Levinas, 2019) is a matter of some debate (see Atterton, 2011; D. Clark, 1997; Derrida, 2008).

9 To be clear, other people are certainly (and routinely) persecuted, tyrannized, and afflicted. I am not denying this. The point is that such forms of violence can never fully subjugate the other. The other always withdraws from tyranny's grasp. As Paul Harrison suggests (personal correspondence, 2022), "when tyranny aims to strike the other, it always misses. But it does hit something."

10 Marilyn Strathern (2018) comes very close to this position when she explores how the 'not-knowing' of others solicits ethnographic attention and thus ethnographic relations. In other words, our relations with others are an effect of a certain distance or non-correspondence in terms of the way the "I" and the "you" understand the relation between us.

11 The full quote is as follows: "The closedness of the separated being must be ambiguous enough for, on the one hand … the interior being to be pursued in an egoist atheism refuted by nothing exterior … but on the other hand *within the very interiority* … there must be produced a heteronomy that incites to another destiny … a shock must be produced which, without inverting the movement of interiorisation, without breaking the thread of interior substance, would furnish occasion for a resumption of relations with exteriority. Interiority must be at the same time closed an open" (Levinas, 1969, pp. 148–9, emphasis in original).

12 This is a tricky one as the term *representational* has, over the last four decades, become associated with what is often termed the *politics of representation*. As we will see, in this chapter and subsequent ones, representation means something more than cognitive representational practices.

13 This other that precedes the subject is referred to by Levinas as the feminine other, a terminology that has been a matter of some controversy (Bergo, 2003; Chanter, 2001; Katz, 2003). But looking beyond the nomenclature, we can take the point that as a site of self-possession, the dwelling must be established as a place apart, and thus, already taking shape in relation to something else. This is Levinas's way of signalling that the other was always already there, before the subject and before recollection.

14 To return to the mythic quality of Levinas's subject, there is something very Eden-like about the journey. As I discuss in the next section's introduction, Adam and Eve are given their enjoyments in Eden, but they do not properly possess them, that is, they do not worry about the morrow. It is only in awakening to their situation (via the tree) that they reckon with the problem of work and toil, that is, of needing shelter and planning for the future.

15 Again, I am unsure to what extent the distinction Levinas is making is useful, as surely animals who nest and den are also engaged in an economy and also planning for the future.

16 To say that the other has no force or capacity is potentially misleading. Better to think of it as a subtractive power, a powerless power, which works to take away from any and every representational economy.

17 While *Being and Time* primarily understands language as following the contours of being-in-the-world, Polt (2013) suggests that Heidegger (even at this early stage) points to how language can break through the everyday: "*Being and Time*'s brief but crucial exploration of the ontological basis of authentic politics … proposes that the people must discover the destiny of a new generation through 'communicating' and 'struggling' (SZ 384). Such communication could not be idle talk; it would have to break new ground" (p. 65).

18 For Heidegger, being-in-the-world is ontological. This correlation will provide the basis for many of Levinas's arguments in *Otherwise than Being, or Beyond Essence*.

19 One can hear echoes of the feminine other here as the one who came before, sets the table, and thus allows a certain modality of subjectivity to take shape.

20 I am reminded here of Lisa Feldman Barrett's (2018) book *How Emotions Are Made*. Her central argument is that emotions are cultural in the sense that they are learned responses to affective experiences. What makes her

analysis interesting is its capacity to illustrate how affective corporeal states are limited by their naming. For Barrett, the various affective states that are currently classified under the terminology of anxiety provides both a basis for comprehension even as it circumscribes and delimits their experience. In other words, the act of naming an affective response "anxiety" makes the sensation both more comprehensible (because it is named) and less comprehensible (because it is never contained by that naming). The thesis has obvious overlaps with Derrida and Lacan.

21 While I take the point that there are many potential worlds and many ways that subjects can see, speak, and conceive what is real within them, for Levinas these worlds are not primary. They represent the multitudinous manner that language allows subjects to distil the saying.

22 By economy I mean a realm of calculated (and calculable) relations.

23 In saying this I in no way mean to diminish the violence such claims can perpetrate. Without question, there is a politics of claiming and violence that is a well-established tool for making one's claims appear real.

Section 3. Geography

1 In saying that exile is tragic I mean this in an emic sense, that is, the story is often told as one of tragedy and abandonment. Many Talmudic commentators, including Levinas, see this ostensible tragedy as the dawn of consciousness and care for the other.

2 The Talmud characterizes this ambivalence as a problem of standing. As Zornberg (1995) suggests, before the fall, Adam and Eve stood in the presence of God, unaware and unconcerned with their position. In exile, however, they lose this certainty, a loss represented in terms of their capacity to stand, "to be banished from the Garden is to lose a particular standing ground" (p. 21). This is what knowledge is – not a sense of being self-assured but a sense of being unself-assured. Both a form of power and an essential unpowering.

7. Culture as Claiming

1 This is a very different conception of alterity than is normally found in the anthropological literature, where alterity usually refers to the difference of other beings. As should be clear by now, alterity, here, does not simply mean difference but something discordant or unaligned with everyday existence, something that punctuates life, disturbing it (fundamentally and unalterably) from another dimension (Levinas, 1987).

2 In contrast to Heidegger who conceptualized inter-world understanding through the concept of hermeneutics.

3 In Heidegger's terms, how would the object appear within Dasein's world?
4 It is this orientation towards the constitutent parts that often lead critics to suggest that Latour's metaphysics is more method than philosophy. While I would not necessarily agree, I understand the sentiment. For Latour, an object is thought primarily in terms of the unity one wants to examine and disaggregate, following the various elements and forces coming together to make that object's presence and appearance – its forcefulness in a world – possible.
5 For Hannah (2019), direction is an underrecognized primordial structure of everyday existence. No being, human or otherwise, can attend to several things at once despite the fact that the world itself demands attention from many directions at the same time.
6 This is a trauma similar to the one Laplanche (1999) described in chapter 4. There is the wound of the event itself (an earthquake) and the wound that precedes the event, the wound that comes from the enigmatic. Like Laplanche, Derrida argues that trauma is predicated upon a reckoning with our primordial exposure to a time we cannot see; a possibility that is always already possible even as it is unknowable.
7 Even referring to the wound as a potential imbues it with a problematic sense of positivity, akin to Deleuze's (1994) notion of the virtual. The wound I am trying to outline here is one that challenges language as it (by definition) defies definition, naming or, indeed, any positive attribution (Harrison, 2021).
8 To be clear, those living at the margins are always more aware of their vulnerability (see Auyero, 2012; Brice, 2020; Coleman, 2016; Hitchen, 2016; Joronen, 2017).
9 In *Of Grammatology*, Derrida (1976) makes a similar point about writing in his discussion of Phaedrus. Writing for Socrates is not a gift of memory but a dangerous supplement that leads to misapprehension, divorced as it is from the immediacy of the dialogic context in which thinking emerges. Thus, memory (for Socrates and the Ok people) emerges as an aporatic problem that writing (or ritual borrowing) seeks to (always problematically) resolve.
10 Jackson (2013a) does not lay out his ideas as programmatically as I am suggesting here so my interpretation is contestable.

8. Fundamental Geography

1 It is important to acknowledge that Material Culture Studies has long argued that material culture is the means by which identities actualize. Daniel Miller (1994, 1998, 2005), in particular, argues that consumption allows subjects to become unalienated from the identities they are given.

While I see this work as essentially dialectical (for Miller alienation is immanent to social relations themselves), there are a number of touch points, some of which I explore in a previous article where a very underdeveloped version of the ideas presented in this book were debated (D. Miller & Tilley, 2011; Rose, 2011a, 2011b).

2 This concept of fabrication shares much in common with Heidegger's notion of *Poiesis* (see Joronen, 2012; Ziarek, 1999).

3 I recognize that I am skimming over a more comprehensive discussion of Lacan's (2008) conception of the unconscious; a concept that requires some attention to do properly but will distract from the main thread of argument. The issue surrounds the kind of knowing that the unconscious represents. For Lacan, the unconscious is a knowing that the subject cannot bear. Thus for Lacan, the inherent negativity of desire is something the unconscious knows but is a knowledge that cannot be tolerated or accepted, what Felman (1987) calls "a kind of unmeant knowledge that escapes intentionality and meaning … [a knowledge] the subject cannot recognise, assume as his, appropriate; a speaking knowledge nonetheless denied to the speaker's knowledge" (p. 77; see also Fink, 1997). When I say that subjects are "concerned" with their desire, I do not mean this in a conscious manner.

4 My use of Lacan is admittedly selective (limited primarily to Seminar VII) and will no doubt raise questions by those working within the psychoanalytic tradition, particularly due to its lack of engagement with Seminars X and XI, which offer a more developed conception of *Das Ding* (through Object a) as well as a more intersubjective (i.e., dialectical) conception of subjectivity (vis-à-vis the Big Other) (see Chanter, 1998). There is a long and somewhat fraught line of literature that circuitously works to bring Levinas and Lacan together and Seminar VII is often the focus of this discussion. Some examples I have found particularly fruitful include J. Butler (2005a), Critchley (1998), Ruti (2015), and Harasym's (1998) excellent collection *Levinas and Lacan: The Missed Encounter*.

5 Balaska (2018) provides the following by way of example: "One such example is the infant's hunger, the need for food which gives rise to demand and, through demand, to desire. If hunger is the need for food, then the demand is not only for milk but also for love, for the experience of attention and care that breastfeeding involves, and the desire is the desire of the mother in her idealized form, as the forever satisfying object, as the breast that is always there" (p. 311). For my reading of Lacan (2008), I understand desire to be something that carries the dimension of the impossible, thus the line "desire of the mother in her idealized form" is particularly salient.

6 The period of representational cultural geography can broadly be conceived as taking a realist approach to representation. By this I mean that

knowledge was generally thought to be divided into two levels. On the one hand, there is the real: knowledge that exists outside the human and was described through ontological statements about what was existent and true. And on the other hand, there is the representational: knowledge that exists within the human realm, was inflected with culture, history, and language and described through epistemological statements about what was knowable. Without question, geographers fell within various camps along this realist divide. Some argued that representation made all ontological statements epistemological (as they were always inflected by culture) and others argued for partial truths, where some real-world knowledge could be gleaned through proper method. The position being developed here is still a form of realism but what I termed in the previous chapter a "negative realism" where the Real is not a positive ontology of being (of what is) but an infinite negativity that subtracts from every positive assertion of truth.

7 In the previous chapter I characterized this framework as negative realism where what is real is existential alterity and where what is representation is ontology.

8 Žižek calls the Jew a master-signifier, modes of signification and debate that work to perpetuate and hide the work of the onto-symbolic order. A good illustration of such master-signifiers can be seen in Žižek's discussion of the film *Jaws* where he explores the various social commentaries about what the shark in the film *means* (voracious capitalism, nature unchained, underage sexuality, etc.). His interest here is how the discussion itself both perpetuates and masks the onto-symbolic playing field upon which such interpretations are already at work: "The analyst already has something to say about society (some point to make about the environment, sexuality or capitalism), which they then attribute to the shark … what is not questioned – what the overwhelming physical presence of the shark allows us to forget – is that this is only an interpretation of society. What is not seen is that circularity according to which the shark is seen as embodying certain tendencies that have already been attributed to the shark" (R. Butler, 2005, p. 45; see also Žižek, 2008a).

9 Both Clarke (2011) and Strohmayer (2021) make similar claims about the presence of landscape being a response to a more primordial absence, though they do so in response to quite different problems.

10 As Clarke (2011) suggests, the literal meaning of production is to "render visible" (p. 952).

11 It is important to recognize that claims can claim in many languages. Even when the claims that subjects make take shape through a language of non-possession (as in the case with many Indigenous ontologies), I would argue that they are still claims in the sense that they articulate a modality of being that is their own.

9. Dreams of Presence

1 These stories come from ethnographic fieldwork I conducted in Egypt in 2000 on the politics of historic preservation.

2 My interpretation of this sign is that it is in relation to litter and pollution rather than personal hygiene.

3 This is not to suggest that I did not try. Just the opposite – it is precisely the mystery of difference, that is, its defiance of coherent explanation – that persistently solicits our attention. Indeed, while there is a long history of geographers arguing that the desire to contemplate, name, and represent difference is an expression of our imperial eyes (McClintock, 1995; Pratt, 1992) – an imperialist desire buried deep within the Western psyche – I would proffer a far more prosaic explanation. It is the defiance of difference (its resistance to coherent explanation) that compels humans to contemplate it.

4 Materiality has taken on an extremely important role in contemporary social theory, particularly in its Deleuzian and Latourian incarnations. In the work of Ingold, Malafouris, Wylie, B. Anderson, and others, objects are seen to be a key element for illustrating how subjectivity takes shape through a range of affective material relationships. While this work is interesting and compelling in its own right, it is not particularly interested in (or oriented towards) the question of culture.

References

Abu-Lughod, L. (1991). Writing against culture. In R.G. Fox (Ed.), *Recapturing anthropology: Working in the present* (pp. 137–62). School of American Research Press.

Adey, P., & Anderson, B. (2011). Event and anticipation: UK civil contingencies and the space-times of decision. *Environment and Planning A, 43*, 2878–99. https://doi.org/10.1068/a43576.

Agnew, J.A., & Livingstone, D.N. (2011). *The SAGE handbook of geographical knowledge*. Sage.

Alberti, B., Fowles, S., Holbraad, M., Marshall, Y., & Witmore, C. (2011). "Worlds otherwise": Archaeology, anthropology, and ontological difference. *Current Anthropology, 52*(6), 896–912. https://doi.org/10.1086/662027.

Althusser, L. (1990). *For Marx*. Verso.

Althusser, L., & Balibar, E.T. (1977). *Reading Capital* (2nd ed.). NLB.

Anderson, B. (1983). *Imagined communities: Reflections on the origin and spread of nationalism*. Verso.

Anderson, B. (2010). Security and the future: Anticipating the event of terror. *Geoforum, 41*(2), 227–35. https://doi.org/10.1016/j.geoforum.2009.11.002.

Anderson, B. (2011). Affect and biopower: Towards a politics of life. *Transactions of the Institute of British Geographers, 36*. https://doi.org/10.1111/j.1475-5661.2011.00441.x.

Anderson, B. (2014). *Encountering affect: Capacities, apparatuses, conditions*. Ashgate.

Anderson, B. (2016). Neoliberal affects. *Progress in Human Geography, 40*(6), 734–53. https://doi.org/10.1177/0309132515613167.

Anderson, B. (2019). Cultural geography III: The concept of "culture." *Progress in Human Geography, 44*(3), 608–17.

Anderson, B., & Harrison, P. (Eds.). (2010). *Taking-place: Non-representational theories and human geography*. Ashgate.

Anderson, B., & McFarlane, C. (2011). Assemblage and geography. *Area*, *43*(2), 124–7. https://doi.org/10.1111/j.1475-4762.2011.01004.x.
Anderson, K., Domash, M., Pile, S., & Thrift, N. (Eds.). (2003). *Handbook of cultural geography*. Sage.
Archer, K. (1993). Regions as social organisms: The Lamarckian characteristics of Vidal de la Blache's regional geography. *Annals of the Association of American Geographers*, *83*(3), 498–514. https://doi.org/10.1111/j.1467-8306.1993.tb01947.x.
Archer, M.S. (1996). *Culture and agency: The place of culture in social theory*. Cambridge University Press.
Argyrou, V. (2002). *Anthropology and the will to meaning: A postcolonial critique*. Pluto Press.
Åsberg, C. (2010). Enter cyborg: Tracing the historiography and ontological turn of feminist technoscience studies. *International Journal of Feminist Technoscience*, *1*, 1–25. https://doi.org/10.1177/1350506810377692.
Ash, J., & Simpson, P. (2016). Geography and post-phenomenology. *Progress in Human Geography*, *40*(1), 48–66. https://doi.org/10.1177/0309132514544806.
Atterton, P. (2011). Levinas and our moral responsibility toward animals. *Inquiry*, *54*(6), 633–49. https://doi.org/10.1080/0020174X.2011.628186.
Auyero, J. (2012). *Patients of the state: The politics of waiting in Argentina*. Duke University Press.
Avins, J. (2015). The dos and don'ts of cultural appropriation. *The Atlantic*. https://www.theatlantic.com/entertainment/archive/2015/10/the-dos-and-donts-of-cultural-appropriation/411292/.
Baker, J.O., Perry, S.L., & Whitehead, A.L. (2020). Crusading for moral authority: Christian nationalism and opposition to science. *Sociological Forum*, *35*(3), 587–607. https://doi.org/10.1111/socf.12619.
Balaska, M. (2018). Can there be happiness in psychoanalysis? Creon and Antigone in Lacan's Seminar VII. *College Literature*, *45*(2), 308–29. https://doi.org/10.1353/lit.2018.0019.
Banet-Weiser, S., & Miltner, K.M. (2016). #MasculinitySoFragile: Culture, structure, and networked misogyny. *Feminist Media Studies*, *16*(1), 171–4. https://doi.org/10.1080/14680777.2016.1120490.
Barker, A.J., & Pickerill, J. (2020). Doings with the land and sea: Decolonising geographies, Indigeneity, and enacting place-agency. *Progress in Human Geography*, *44*(4), 640–62. https://doi.org/10.1177/0309132519839863.
Barnett, C. (2001). Culture, geography and the arts of government. *Environment and Planning D: Society and Space*, *19*, 7–24. https://doi.org/10.1068/d236.
Barnett, C. (2009). Culture. In D. Gregory, R. Johnston, G. Pratt, M. Watts, & S. Whatmore (Eds.), *The dictionary of human geography* (pp. 135–8). Wylie.
Barrett, L.F. (2018). *How emotions are made: The secret life of the brain*. Pan Books.
Barthes, R. (1977). *Image, music, text* (S. Heath, Trans.). Fontana.

Bataille, G. (1985). *Visions of excess: Selected writings, 1927–1939* (A. Stoekl, C. Lovitt, & D.L. Jr., Trans.). University of Minnesota Press.

Battaglia, D. (1999). Towards an ethics of the open subject: Writing culture in good conscience. In H.L. Moore (Ed.), *Anthropological theory today* (pp. 115–49). Blackwell Publishers.

Baudrillard, J. (1989). *America* (C. Turner, Trans.). Verso.

Behar, R., & Gordon, D.A. (1995). *Women writing culture*. University of California Press.

Bender, B., & Aitken, P. (1998). *Stonehenge: Making space*. Berg.

Bennett, J. (2010). *Vibrant matter: A political ecology of things*. Duke University Press.

Bennett, T. (2013). Habit: Time, freedom, governance. *Body and Society, 19*(2&3), 107–35. https://doi.org/10.1177/1357034X13475828.

Bennett, T. (2015). Cultural studies and the culture concept. *Cultural Studies, 29*(4), 546–68. https://doi.org/10.1080/09502386.2014.1000605.

Berger, J. (1972a). *Selected essays and articles: The look of things*. Pelican.

Berger, J. (1972b). *Ways of seeing*. Harmondsworth.

Bergo, B. (2003). *Levinas between ethics and politics: For the beauty that adorns the earth*. Duquesne University Press.

Bergson, H. (1959). *Time and free will: An essay on the immediate data of consciousness*. G. Allen & Unwin.

Bissell, D. (2009). Travelling vulnerabilities: Mobile timespace of quiescence. *Cultural Geographies, 16*, 427–45. https://doi.org/10.1177/1474474009340086.

Bissell, D. (2012). Agitating the powers of habit: Towards a volatile politics of thought. *Theory and Event, 15*(1). http://muse.jhu.edu/journals/theory_and_event/v015/015.011.bissell.html.

Bissell, D. (2013). Habit displaced: The disruption of skillful performance. *Geographical Research, 51*(2), 120–9. https://doi.org/10.1111/j.1745-5871.2012.00765.x.

Bissell, D. (2015). Virtual infrastructures of habit: The changing intensities of habit through gracefulness, restlessness and clumsiness. *Cultural Geographies, 22*(1), 127–46. https://doi.org/10.1177/1474474013482812.

Bissell, D. (2016). Micropolitics of mobility: Public transport commuting and everyday encounters with forces of enablement and constraint. *Annals of the Association of American Geographers, 106*(2), 394–403. https://doi.org/10.1080/00045608.2015.1100057.

Bissell, D. (2022). The anaesthetic politics of being unaffected: Embodying insecure digital platform labour. *Antipode, 54*(1), 86–105. https://doi.org/10.1111/anti.12786.

Bissell, D. (2023). Negative urbanism: Unknowability, illegibility and ambivalence in the platform city. *City, 27*(1–2), 56–75. https://doi.org/10.1080/13604813.2022.2145633.

Bissell, D., Rose, M., & Harrison, P. (2021). *Negative geographies: Exploring the politics of limits*. University of Nebraska Press.

Blaser, M. (2009). Political ontology: Cultural studies without "cultures"? *Cultural Studies, 23*(5), 873–96. https://doi.org/10.1080/09502380903208023.

Blaser, M. (2014). Ontology and indigeneity: On the political ontology of heterogeneous assemblages. *Cultural Geographies, 21*(1), 49–58. https://doi.org/10.1177/147447401246253.

Bloch, M. (2005). Where did anthropology go?: Or the need for "human nature." In M. Bloch (Ed.), *Essays on cultural transmission. LSE monographs on social anthropology* (pp. 1–20). Berg.

Blum, B. (2020). The radical history of corporate sensitivity training. *The New Yorker*. https://www.newyorker.com/culture/cultural-comment/the-radical-history-of-corporate-sensitivity-training.

Blumenfeld, J. (2014). Egoism, labour, and possession: A reading of "Interiority and Economy," Section II of Lévinas' *Totality of Infinity*. *Journal of the British Society for Phenomenology, 45*(2), 107–17. https://doi.org/10.1080/00071773.2014.919120.

Boas, F. (1940). *Race, language and culture*. The Macmillan Company.

Bonds, A. (2020). Race and ethnicity II: White women and the possessive geographies of white supremacy. *Progress in Human Geography, 44*(4), 778–88. https://doi.org/10.1177/0309132519863479.

Bourdieu, P. (1990). *The logic of practice*. Stanford University Press.

Braun, B. (2008). Environmental issues: Inventive life. *Progress in Human Geography, 32*(5), 667–79. https://doi.org/10.1177/0309132507088030.

Brice, S. (2020). Geographies of vulnerability: Mapping transindividual geometries of identity and resistance. *Transactions of the Institute of British Geographers, 45*(6), 664–77. https://doi.org/10.1111/tran.12358.

Brody, D. (1998). Levinas and Lacan: Facing the Real. In S. Harasym (Ed.), *Levinas and Lacan: The missed encounter* (pp. 56–78). State University of New York Press.

Brooks, A.C. (2015). The real victims of victimhood. *The New York Times*. https://www.nytimes.com/2015/12/27/opinion/sunday/the-real-victims-of-victimhood.html?searchResultPosition=8.

Brown, W. (2006). *Regulating aversion: Tolerance in the age of identity and empire*. Princeton University Press.

Brown, W., Forst, R., Holzhey, C.F.E., & Di Blasi, L. (2014). *The power of tolerance: A debate*. Columbia University Press.

Butler, J. (1997). *The psychic life of power: Theories in subjection*. Stanford University Press.

Butler, J. (2004). *Precarious life: The powers of mourning and violence*. Verso.

Butler, J. (2005). *Giving an account of oneself*. Fordham University Press.

Butler, R. (2005). *Slavoj Žižek: Live theory*. Continuum.

Buttimer, A. (1971). *Society and milieu in the French geographic tradition*. Rand McNally.

Buttimer, A. (1976). Grasping the dynamism of the lifeworld. *Annals of the Association of American Geographers*, *66*(2), 277–92. https://doi.org/10.1111/j.1467-8306.1976.tb01090.x.

Butzer, K., & Butzer, E.K. (1993). The sixteenth-century environment of the central Bajio: Archival reconstruction from colonial land grants and the question of Spanish ecological impact. In K. Mathewson (Ed.), *Culture, form, and place: Essays in cultural and historical geography* (Vol. 32, *Geoscience and Man*, pp. 89–124). Geoscience Publications, Department of Geography and Anthropology, Louisiana State University.

Cameron, E., de Leeuw, S., & Desbiens, C. (2014). Indigeneity and ontology. *Cultural Geographies*, *21*(1), 19–26. https://doi.org/10.1177/1474474013500229.

Campbell, R.D. (1968). Personality as an element of regional geography. *Annals of the Association of American Geographers*, *58*(4), 748–59. https://doi.org/10.1111/j.1467-8306.1968.tb01665.x.

Campbell, A., & Strachan, M. (2018). NY Daily News editor acccused of harrasment fostered a culture of silence. *Huffington Post*. https://www.huffpost.com/entry/daily-news-rob-moore_n_5a68d1c8e4b0e5630075d434.

Carlisle, C. (2008). Editor's introduction. In F. Ravaisson (Ed.), *Of habit* (pp. 1–21). Continuum.

Carlisle, C. (2010). Between freedom and necessity: Felix Ravaisson on habit and the moral life. *Inquiry*, *53*(2), 123–45. https://doi.org/10.1080/00201741003612146.

Carlisle, C. (2013). The question of habit in theology and philosophy: From hexis to plasticity. *Body and Society*, *19*(2&3), 30–57. https://doi.org/10.1177/1357034X12474475.

Carrier, J.G. (2003). Mind, gaze and engagement: Understanding the environment. *Journal of Material Culture*, *8*(1), 5–23. https://doi.org/10.1177/1359183503008001760.

Carrithers, M., Candea, M., Sykes, K., & Holbraad, M. (2010). Ontology is just another word for culture: motion tabled at the 2008 meeting of the Group for Debates in Anthropological Theory, University of Manchester. *Critique of Anthropology*, *30*(2), 152–200. https://doi.org/10.1177/0308275X09364070.

Carter, J., & Hollinsworth, D. (2017). Teaching Indigenous geography in a neo-colonial world. *Journal of Geography in Higher Education*, *41*(2), 182–97. https://doi.org/10.1080/03098265.2017.1290591.

Carter-White, R., & Doel, M.A. (2022). The signature of the disaster: Witnessness in death camp and tsunami survivor testimony. *Transactions of the Institute of British Geographers*, *47*(2), 455–69. https://doi.org/10.1111/tran.12511.

Caruth, C. (2014). *Listening to trauma: Conversations with leaders in the theory and treatment of catastrophic experience*. Johns Hopkins University Press.

Cavell, S. (1999). *The claim of reason: Wittgenstein, skepticism, morality, and tragedy* (New ed.). Oxford University Press.

Chandler, D. (2013). International statebuilding and the ideology of resilience. *Politics, 33*(4), 276–86. https://doi.org/10.1111/1467-9256.12009.

Chandler, D., & Pugh, J. (2021). Anthropocene islands: There are only islands after the end of the world. *Dialogues in Human Geography, 11*(3), 395–415. https://doi.org/10.1177/2043820621997018.

Chanter, T. (1998). Reading Hegel as a mediating master: Lacan and Levinas. In S. Harasym (Ed.), *Levinas and Lacan: The missed encounter* (pp. 1–21). State University of New York Press.

Chanter, T. (2001). *Time, death, and the feminine: Levinas with Heidegger*. Stanford University Press.

Clark, A.H. (1954). Historical geography. In P.E. James, C.F. Jones, & J.K. Wright (Eds.), *American geography: Inventory and prospect* (pp. 70–105). Association of American Geographers.

Clark, D. (1997). On being "the last Kantian in Nazi Germany': Dwelling with animals after Levinas. In J. Ham & M. Senior (Eds.), *Animal acts: Configuring the human in western history* (pp. 165–98). Routledge.

Clarke, D.B. (2011). Utopologies. *Environment and Planning D: Society and Space, 29*(6), 951–67. https://doi.org/10.1068/d9210.

Clement, V. (2019). Beyond the sham of the emancipatory Enlightenment: Rethinking the relationship of Indigenous epistemologies, knowledges, and geography through decolonizing paths. *Progress in Human Geography, 43*(2), 276–94. https://doi.org/10.1177/0309132517747315.

Clifford, J. (1983). On ethnographic authority. *Representations, 1*(2), 118–46. https://doi.org/10.2307/2928386.

Clifford, J. (1986). Introduction: Partial truths. In J. Clifford & G.E. Marcus (Eds.), *Writing culture: The poetics and politics of ethnography* (pp. 1–26). University of California Press.

Clifford, J. (1988). *The predicament of culture: Twentieth-century ethnography, literature, and art*. Harvard University Press.

Clifford, J., & Marcus, G.E. (1986). *Writing culture: The poetics and politics of ethnography*. University of California Press.

Cloke, P., & Jones, O. (2001). Dwelling, place, and landscape: An orchard in Somerset. *Environment and Planning A, 33*(4), 649–66. https://doi.org/10.1068/a3383.

Cloke, P., & Jones, O. (2004). Turning in the graveyard: Trees and the hybrid geographies of dwelling, monitoring and resistance in a Bristol cemetery. *Cultural Geographies, 11*, 313–41. https://doi.org/10.1191/1474474004eu300oa.

Cloke, P.J., Philo, C., & Sadler, D. (1991). *Approaching human geography: An introduction to contemporary theoretical debates*. Guilford Press.

Clout, H. (2003). Place description, regional geography and area studies: the chorographic inheritance. In R. Johnston & M. Williams (Eds.), *A century of British geography* (pp. 247–74). British Academy Scholarship.
Cockayne, D. (2018). Underperformative economies: Discrimination and gendered ideas of workplace culture in San Francisco's digital media sector. *Environment and Planning A: Economy and Space, 50*(4), 754–72. https://doi.org/10.1177/0308518X18754883.
Colebrook, C. (2000). From radical representations to corporeal becomings: The feminist philosophy of Lloyd, Grosz, and Gatens. *Hypatia, 15*(2), 76–93. https://doi.org/10.1111/j.1527-2001.2000.tb00315.x.
Coleman, R. (2016). Austerity futures: Debt, temporality and (hopeful) pessimism as an austerity mood. *New Formations, 87*(87), 102–18. https://doi.org/10.3898/NEWF.87.5.2016.
Cosgrove, D. (1983). Towards a radical cultural geography: Problems of theory. *Antipode, 15*, 1–11. https://doi.org/10.1111/j.1467-8330.1983.tb00318.x.
Cosgrove, D. (1985). Prospect, perspective and the evolution of the landscape idea. *Transactions for the Institute of British Geographers, 10*, 45–62. https://doi.org/10.2307/622249.
Country, B., Wright, S., Suchet-Pearson, S., Lloyd, K., Burarrwanga, L., Ganambarr, R., Ganambarr-Stubbs, M., Ganambarr, B., & Maymuru, D. (2015). Working with and learning from Country: Decentring human author-ity. *Cultural Geographies, 22*(2), 269–83. https://doi.org/10.1177/1474474014539248.
Crang, M. (1998). *Cultural geography*. Routledge.
Cresswell, T. (1996). *In place/out of place*. University of Minnesota Press.
Cresswell, T. (2013). *Geographic thought: A critical introduction*. Wiley-Blackwell.
Critchley, S. (1998). Das Ding: Lacan and Levinas. *Research in Phenomenology, 28*(1), 72–90. https://doi.org/10.1163/156916498x00056.
Critchley, S. (1999a). *The ethics of deconstruction: Derrida and Levinas*. Edinburgh University Press.
Critchley, S. (1999b). *Ethics, politics, subjectivity: Essays on Derrida, Levinas and contemporary French thought*. Verso.
Cuttitta, P. (2014). Migration control in the Mediterranean Grenzsaum: Reading Ratzel in the Strait of Sicily. *Journal of Borderlands Studies, 29*(2), 117–31. https://doi.org/10.1080/08865655.2014.915701.
Cvijic, J. (2016). La peninsule Balkanique, geographie humaine *(Éd. 1918)*. Paris: Hachette BNF.
Daniels, S. (1989). Marxism, culture, and the duplicity of landscape. In R. Peet & N. Thrift (Eds.), *New models in geography: The political economy perspective* (pp. 196–220). Routledge.
Davis, J.T., & Perry, S.L. (2020). White Christian nationalism and relative political tolerance for racists. *Social Problems, 68*(3), 513–34. https://doi.org/10.1093/socpro/spaa002.

Daya, S. (2019). Words and worlds: Textual representation and new materialism. *Cultural Geographies*, *26*(3), 361–77. https://doi.org/10.1177/1474474019832356.

De la Cadena, M. (2010). Indigenous cosmopolitics: Conceptual reflections beyond politics as usual. *Cultural Anthropology*, *25*(2), 334–70. https://doi.org/10.1111/j.1548-1360.2010.01061.x.

De la Cadena, M. (2015). *Earth beings: Ecologies of practice across Andean worlds*. Duke University Press.

Dekeyser, T., & Jellis, T. (2021). Besides affirmationism? On geography and negativity. *Area*, *53*(2), 318–25. https://doi.org/10.1111/area.12684.

Dekeyser, T., Secor, A., Rose, M., Bissell, D., Zhang, V., & Romanillos, J. L. (2022). Negativity: Space, politics and affects. *Cultural Geographies*, *29*(1), 5–21. https://doi.org/10.1177/14744740211058080.

Deleuze, G. (1990). *The logic of sense* (M. Lester & C. Stivale, Trans.). Althlone Press.

Deleuze, G. (1991). *Empiricism and subjectivity: An essay on Hume's theory of human nature*. Columbia University Press.

Deleuze, G. (1994). *Difference and repetition* (P. Patton, Trans.). Columbia University Press.

Deleuze, G. (2004). *Desert islands and other texts, 1953–1974*. MIT Press.

Deleuze, G., & Guattari, F. (1987). *A thousand plateaus: Capitalism and schizophrenia* (B. Massumi, Trans.). Athlone.

Deleuze, G., & Guattari, F. (1994). *What is philosophy?* Columbia University Press.

Denevan, W.M. (2001). *Cultivated landscapes of native Amazonia and the Andes*. Oxford University Press.

Derrida, J. (1976). *Of grammatology* (G.C. Spivak, Trans.; 1st American ed.). Johns Hopkins University Press.

Derrida, J. (1978). *Writing and difference* (A. Bass, Trans.). University of Chicago Press.

Derrida, J. (1992). Force of law: "The mystical foundation of authority." In D. Cornell & M. Rosenfield (Eds.), *Deconstruction and the possibility of justice* (pp. 230–98). Routledge.

Derrida, J. (1993). *Given time 1: Counterfeit money*. University of Chicago Press.

Derrida, J. (1997). *Politics of friendship*. Verso.

Derrida, J. (2000). Hospitality. *Angelaki*, *5*(3), 3–18.

Derrida, J. (2001). *On cosmopolitanism and forgiveness*. Routledge.

Derrida, J. (2005a). *Rogues: Two essays on reason*. Stanford University Press.

Derrida, J. (2005b). *Sovereignties in question: The poetics of Paul Celan* (T. Dutoit & O. Pasanen, Trans.). Fordham University Press.

Derrida, J. (2008). *The animal that therefore I am*. Fordham University Press.

Derrida, J. (2009). *The beast and the sovereign, volume 1* (G. Bennington, Trans.). University of Chicago Press.

Descola, P. (2013). *Beyond nature and culture*. Chicago University Press.
Descola, P., & Gaisli, P. (1996). *Nature and society: Anthropological perspectives*. Routledge.
Dewsbury, J.D. (2011). The Deleuze-Guattarian assemblage: Plastic habits. *Area, 43*(2), 148–53. https://doi.org/10.1111/j.1475-4762.2011.01006.x.
Dewsbury, J.D. (2015). Non-representational landscapes and the performative affective forces of habit: From "Live" to "Blank." *Cultural Geographies, 22*(1), 29–47. https://doi.org/10.1177/1474474014561575.
Dewsbury, J.D. (2002). Embodying time, imagined and sensed. *Time and Society, 11*(1), 147–54. https://doi.org/10.1177/0961463X02011001010.
Dewsbury, J.D., Harrison, P., Rose, M., & Wylie, J. (2002). Enacting geographies: Editorial introduction. *Geoforum, 33*(4), 437–40. https://doi.org/10.1016/s0016-7185%2802%2900029-5.
Donato, J. D. (2020). Blackfishing, influencers and cancel culture: A tangled web. *The Huffington Post*. https://www.huffpost.com/entry/blackfishing-cancel-culture_l_5f32dab2c5b64cc99fdefb73.
Dostol, R.J. (1982). The problem of "indifferenz" in *Sein und Zeit*. *Philosophy and Phenomenological Research, 43*, 43–58. https://doi.org/10.2307/2107512.
Douthat, R. (2010). The roots of white anxiety. *The New York Times*. https://www.nytimes.com/2010/07/19/opinion/19douthat.html?searchResultPosition=9.
Douthat, R. (2020). 10 theses about cancel culture. *The New York Times*. https://www.nytimes.com/2020/07/14/opinion/cancel-culture-.html?searchResultPosition=3.
Dreyfus, H.L. (1991). *Being-in-the-world: A commentary on Heidegger's Being and Time, division I*. MIT Press.
Driver, F. (1985). Power space and the body: A critical assessment of Foucault's *Discipline and Punish*. *Environment and Planning D: Society and Space, 3*, 425–46. https://doi.org/10.1068/d030425.
Driver, F. (1992). Geography's empire: Histories of geographical knowledge. *Environment and Planning D: Society and Space, 10*, 23–40. https://doi.org/10.1068/d100023.
Driver, F. (1997). Bodies in space: Foucault's account of disciplinary power. In T. Barnes & D. Gregory (Eds.), *Reading human geography: The poetics and politics of inquiry* (pp. 279–89). Arnold.
Driver, F., & Gilbert, D. (1999). *Imperial cities: Landscape, display and identity*. Manchester University Press.
Dubow, J. (2011). Still-life, after-life, *nature morte*: W.G. Sebald and the demands of landscape. In S. Daniels, D. DeLyser, J.N. Entrikin, & D. Richardson (Eds.), *Envisioning landscapes, making worlds : geography and the humanities* (pp. 188–97). Routledge.
Dubow, J. (2021). *In exile: Geography, philosophy and Judaic thought*. Bloomsbury Academic.

Dulley, I. (2019). *On the emic gesture: Difference and ethnography in Roy Wagner*. Routledge.

Duncan, J., & Duncan, N. (1988). (Re)reading the landscape. *Environment and Planning D: Society and Space, 6*, 117–26. https://doi.org/10.1068/d060117.

Duncan, J.S., & Gregory, D. (1999). *Writes of passage: Reading travel writing*. Routledge.

Dupuis, A., & Thorns, D.C. (1998). Home, home ownership and the search for ontological security. *The Sociological Review, 46*(1), 24–47. https://doi.org/10.1111/1467-954X.00088.

Elden, S. (2001). *Mapping the present: Heidegger, Foucault and the project of spatial history*. Continuum.

Escobar, A. (2007). The "ontological turn" in social theory: A commentary on "Human Geography without Scale," by Sallie Marston, John Paul Jones II and Keith Woodward. *Transactions of the Institute of British Geographers, 32*(1), 106–11. https://doi.org/10.1111/j.1475-5661.2007.00243.x.

Evens, T. (2008). *Anthropology as ethics: Nondualism and the conduct of sacrifice*. Berg.

Feely, M. (2016). Disability studies after the ontological turn: A return to the material world and material bodies without a return to essentialism. *Disability & Society, 31*(7), 863–83. https://doi.org/10.1080/09687599.2016.1208603.

Felman, S. (1987). *Jacques Lacan and the adventure of insight: Psychoanalysis in contemporary culture*. Harvard University Press.

Fink, B. (1997). *A clinical introduction to Lacanian psychoanalysis: Theory and technique*. Harvard University Press.

Fletcher, J. (1999). Introduction: Psychoanalysis and the question of the other. In J. Laplanche, *Essays on otherness* (pp. 1–51). Routledge.

Forman, P. (2021). Materialist dialogues and the granular. *Dialogues in Human Geography, 11*(2), 307–10. https://doi.org/10.1177/20438206211004859.

Forster, S. (2002). War commemoration and the republic in crisis: Weimar Germany and the Neue Wache. *Central European History, 35*(4). https://doi.org/10.1163/156916102770891179.

Foucault, M. (1972a). *Archaeology of knowledge and the discourse on language* (A.M.S. Smith, Trans.). Pantheon Books.

Foucault, M. (1972b). *Power/knowledge: Selected interviews and other writings 1972–1977* (C. Gordon, Trans.). Pantheon Books.

Foucault, M. (1986). Of other spaces. *Diacritics, 16*(1), 22–7. https://doi.org/10.2307/464648.

Foucault, M. (1995). *Discipline and punish: The birth of the prison*. Vintage.

Foucault, M. (2003). *"Society must be defended": Lectures at the Colláege de France, 1975–76*. Picador.

Foucault, M. (2007). *Security, territory, population: Lectures at the Colláege de France, 1977–78*. Palgrave Macmillan.

Fox, R.G., & King, B.J. (2002). *Anthropology beyond culture*. Berg.

Freud, S. (1955). Beyond the pleasure principle. In J. Strachey & A. Freud (Eds.), *The standard edition of the complete psychological works of Sigmund Freud Volume XVIII (1920–1922): Beyond the pleasure principle – group psychology & other works*. Psychoanalytic Electronic Publishing. https://pep-web.org/browse/document/SE.018.0000A?page=PR0004.

Friedersdorf, C. (2015a). A food fight at Oberlin College. *The Atlantic*. https://www.theatlantic.com/politics/archive/2015/12/the-food-fight-at-oberlin-college/421401/.

Friedersdorf, C. (2015b). The rise of victimhood culture. *The Atlantic*. https://www.theatlantic.com/politics/archive/2015/09/the-rise-of-victimhood-culture/404794/.

Friedersdorf, C. (2017). The destructiveness of call-out culture on campus. *The Atlantic*. https://www.theatlantic.com/politics/archive/2017/05/call-out-culture-is-stressing-out-college-students/524679/.

Frum, D. (2018). Every culture appropriates. *The Atlantic*. https://www.theatlantic.com/ideas/archive/2018/05/cultural-appropriation/559802/.

Gandy, M., & Jasper, S. (2017). Geography, materialism, and the neo-vitalist turn. *Dialogues in Human Geography*, *7*(2), 140–4. https://doi.org/10.1177/2043820617717848.

Geertz, C. (1973). *The interpretation of cultures: Selected essays*. Basic Books.

Geertz, C. (1983). Local knowledge: Fact and law in comparative perspective. In C. Geertz (Ed.), *Local knowledge: Further essays in interpretive anthropology* (pp. 167–234). Basic Books.

Geertz, C. (1995). *After the fact: Two countries, four decades, one anthropologist*. Harvard University Press.

Gibson, C. (2017). Culture. In D. Richardson, N. Castree, M.F. Goodchild, A. Kobayashi, W. Liu, & R.A. Marston (Eds.), *International encyclopedia of geography: People, the earth, environment and technology* (pp. 1–4). Wylie.

Goffman, E. (1959). *The presentation of self in everyday life*. Anchor.

Goldberg, J. (2016, April). The Obama doctrine: The U.S. president talks through his hardest decisions about America's role in the world. *The Atlantic*. https://www.theatlantic.com/magazine/archive/2016/04/the-obama-doctrine/471525/.

Goldberg, R.A. (2001). *Enemies within: The culture of conspiracy in modern America*. Yale University Press.

Gow, P. (1995). Land, people, and paper in Western Amazonia. In E. Hirsch & M. O'Hanlon (Eds.), *The anthropology of landscape* (pp. 43–62). Clarendon Press.

Graeber, D. (2015). Radical alterity is just another way of saying "reality": A reply to Eduardo Viveiros de Castro. *HAU: Journal of Ethnographic Theory*, *5*(2), 1–41. https://doi.org/10.14318/hau5.2.003.

Gramsci, A. (1995). *Selections from the prison notebooks*. International Publishers.

Green, E. (2017). It was culture anxiety that drove white, working-class voters to Trump. *The Atlantic*. https://www.theatlantic.com/politics/archive/2017/05/white-working-class-trump-cultural-anxiety/525771/.
Greg, M. (2015). How sexism shaped corporate culture. *The Atlantic*. https://www.theatlantic.com/business/archive/2015/09/sexism-corporate-culture/407260/.
Gregory, D. (1994). *Geographic imaginations*. Blackwell.
Gregson, N. (2007). *Living with things: Ridding, accommodation, dwelling*. Sean Kingston Publishing.
Grossberg, L. (2006). Does cultural studies have futures? Should it? (Or what's the matter with New York?). *Cultural Studies, 20*(1), 1–32. https://doi.org/10.1080/09502380500492541.
Grosz, E. (1994). *Volatile bodies: Toward a corporeal feminism*. Allen and Unwin.
Gruffudd, P. (1995). Remaking Wales: Nation-building and the geographical imagination, 1925–1950. *Political Geography, 14*(3), 219–39. https://doi.org/10.1016/0962-6298(95)93185-L.
Gupta, A., & Ferguson, J. (1997). *Beyond "culture": Space, identity, and the politics of difference*. Duke University Press.
Haar, M. (1993). *The song of the earth: Heidegger and the grounds of the history of being*. Indiana University Press.
Hannah, M.G. (1993). Foucault on theorizing specificity. *Environment and Planning D: Society and Space, 11*, 349–63. https://doi.org/10.1068/d110349.
Hannah, M.G. (2000). *Governmentality and the mastery of territory in nineteenth-century America*. Cambridge University Press.
Hannah, M.G. (2016). Formations of "Foucault" in Anglo-American geography: An archaeological sketch. In S. Elden & J.W. Crampton (Eds.), *Space, knowledge and power: Foucault and geography*. Taylor & Francis Group.
Hannah, M.G. (2019). *Direction and socio-spatial theory: A political economy of oriented practice*. Routledge.
Harasym, S. (1998). *Levinas and Lacan: The missed encounter*. State University of New York Press.
Haraway, D.J. (2003). *The companion species manifesto: Dogs, people, and significant otherness*. Prickly Paradigm Press.
Haraway, D.J. (2008). *When species meet*. University of Minnesota Press.
Haraway, D.J. (2016). *Staying with the trouble: Making kin in the Chthulucene*. Duke University Press.
Harman, G. (2002). *Tool-being: Heidegger and the metaphyics of objects*. Open Court.
Harman, G. (2009). Dwelling with the fourfold. *Space and Culture, 12*, 292–302. https://doi.org/10.1177/1206331209337080.
Harrison, P. (2007a). How shall I say it…? Relating the non-relational. *Environment and Planning A, 39*(3), 590–608. https://doi.org/10.1068/a3825.

Harrison, P. (2007b). The space between us: Opening remarks on the concept of dwelling. *Environment and Planning D: Society and Space, 25*(4), 625–47. https://doi.org/10.1068/d365t.

Harrison, P. (2008). Corporeal remains: Vulnerability, proximity, and living on after the end of the world. *Environment and Planning A, 40*(2), 423–45. https://doi.org/10.1068/a391.

Harrison, P. (2009). In the absence of practice. *Environment and Planning D: Society and Space, 27*(6), 987–1009. https://doi.org/10.1068/d7907.

Harrison, P. (2015). After affirmation, or, being a loser: On vitalism, sacrifice and cinders. *GeoHumanities, 1*, 285–306. https://doi.org/10.1080/2373566X.2015.1109469.

Harrison, P. (2021). A love whereof non-shall speak: Reflections on naming; of "non-representational theory." In D. Bissell, M. Rose, & P. Harrison (Eds.), *Negative geographies* (pp. 67–91). University of Nebraska Press.

Hart, J.F. (1982). The highest form of the geographer's art. *Annals of the Association of American Geographers, 72*(1), 1–29. https://doi.org/10.1111/j.1467-8306.1982.tb01380.x.

Hartshorne, R. (1939). *The nature of geography: A critical survey of current thought in the light of the past*. Association of American Geographers.

Heidegger, M. (1959). *An introduction to metaphysics* (R. Manheim, Trans.). Yale University Press.

Heidegger, M. (1962). *Being and time* (J. Macquarrie & E. Robinson, Trans.). Harper and Row.

Heidegger, M. (1971a). Building dwelling thinking (A. Hofstadter, Trans.). In *Poetry, language, thought* (pp. 143–62). Harper and Row.

Heidegger, M. (1971b). Language (A. Hofstadter, Trans.). In *Poetry, language, thought* (pp. 187–210). Harper and Row.

Heidegger, M. (1971c). The origin of the work of art (A. Hofstadter, Trans.). In *Poetry, language, thought* (pp. 15–88). Harper and Row.

Heidegger, M. (1971d). "… Poetically man dwells …" (A. Hofstadter, Trans.). In *Poetry, language, thought* (pp. 211–29). Harper and Row.

Heidegger, M. (1971e). The thing (A. Hofstadter, Trans.). In *Poetry, language, thought* (pp. 163–86). Harper and Row.

Heidegger, M. (1977a). The question concerning technology (W. Lovitt, Trans.). In *The question concerning technology, and other essays* (pp. 3–35). Harper and Row.

Heidegger, M. (1977b). The turning (W. Lovitt, Trans.). In *The question concerning technology, and other essays* (pp. 36–49). Harper and Row.

Heidegger, M. (1978a). Letter on humanism. In D.F. Krell (Ed.), *Basic writings from "Being and time" (1927) to "The task of thinking" (1964)* (pp. 189–242). Routledge and Kegan Paul.

Heidegger, M. (1978b). What calls for thinking. In D.F. Krell (Ed.), *Basic writings from "Being and time" (1927) to "The task of thinking" (1964)* (pp. 341–68). Routledge and Kegan Paul.

Heidegger, M. (1995). *The fundamental concepts of metaphysics: World, finitude, solitude* (W. McNeill & N. Walker, Trans.). Indiana University Press.

Heidegger, M. (1996). *Being and time: A translation of Sein und Zeit* (J. Stambaugh, Trans.). State University of New York Press.

Heidegger, M. (1998). On the essence and concept of *physis* in Aristotle's Physics B, I (T. Sheehan, Trans.). In W. McNeill (Ed.), *Pathmarks* (pp. 183–230). Cambridge University Press.

Heidegger, M. (2000). *Elucidations of Holderlin's poetry* (K. Hoeller, Trans.). Humanity Books.

Heidegger, M. (2010). *Country path conversations* (English ed.). Indiana University Press.

Hemmings, C. (2005). Invoking affect. *Cultural Studies, 19*(5), 548–67. https://doi.org/10.1080/09502380500365473.

Henare, A.J.M., Holbraad, M., & Wastell, S. (2007). Introduction: Thinking through things. In A.J.M. Henare, M. Holbraad, & S. Wastell (Eds.), *Thinking through things: Theorising artefacts ethnographically* (pp. 1–31). Routledge.

Hinchliffe, S. (2008). Reconstituting nature conservation: towards a careful political ecology. *Geoforum, 39*, 88–97. https://doi.org/10.1016/j.geoforum.2006.09.007.

Hitchen, E. (2016). Living and feeling the austere. *New Formations, 87*(87), 102–18. https://doi.org/10.3898/NEWF.87.6.2016.

Hobbes, M. (2020). Don't fall for the "cancel culture" scam. *Huffington Post*. https://www.huffpost.com/entry/cancel-culture-harpers-jk-rowling-scam_n_5f0887b4c5b67a80bc06c95e.

Holbraad, M. (2007). The power of powder: Multiplicity and motion in the divinatory cosmology of Cuban Ifa (or mana, again). In A. Henare, M. Holbraad, & S. Wastell (Eds.), *Thinking through things: Theorising artefacts ethnographically* (pp. 189–225). Routledge.

Holbraad, M., & Pedersen, M.A. (2017). *The ontological turn: An anthropological exposition*. Cambridge University Press.

Horton, J., & Kraftl, P. (2014). *Cultural geographies: An introduction*. Routledge.

Hu, J. (2020). The new "Mulan's" uncomfortable relationship with China's past and present. *The New Yorker*. https://www.newyorker.com/culture/cultural-comment/the-new-mulanandnbsps-uncomfortable-relationship-with-chinas-past-and-present.

Hunt, S. (2014). Ontologies of Indigeneity: The politics of embodying a concept. *Cultural Geographies, 21*(1), 27–32. https://doi.org/10.1177/1474474013500226.

Hunter, J.M. (1983). *Perspective on Ratzel's political geography*. University Press of America.

Huntington, E. (1934). *Principles of human geography. Fourth edition, rewritten*. J. Wiley & Sons.

Hutchens, B.C. (2004). *Levinas: A guide for the perplexed*. Continuum.

Imort, M. (2000). *Forestopia: The use of the forest landscape in naturalising national socialist ideologies of volk, race and Lebensraum 1918–1945* (Publication Number 0-612-54061-8) [Doctoral dissertation, Queen's University].

Ingold, T. (1976). *The Skolt Lapps today*. Cambridge University Press.

Ingold, T. (1980). *Hunters, pastoralists, and ranchers: Reindeer economies and their transformations*. Cambridge University Press.

Ingold, T. (2000). *The perception of the environment: Essays in livelihood, dwelling and skill*. Routledge.

Ingold, T. (2007). *Lines: A brief history*. Routledge.

Ingold, T. (2008). When ANT meets SPIDER: Social theory for arthropods. In C. Knappett & L. Malafouris (Eds.), *Material agency: Towards a non-anthropocentric approach* (pp. 209–15). Springer.

Ingold, T. (2011). *Being alive: Essays on movement, knowledge and description*. Routledge.

Ingold, T. (2012). Toward an ecology of materials. *The Annual Review of Anthropology, 2012*(41), 427–42. https://doi.org/10.1146/annurev-anthro-081309-145920.

Ioris, A.A., Benites, T., & Goettert, J.D. (2019). Challenges and contribution of Indigenous geography: Learning with and for the Kaiowa-Guarani of South America. *Geoforum, 102*, 137–41. https://doi.org/10.1016/j.geoforum.2019.03.023.

Jabès, E. (1989). *The book of shares* (R. Waldrop, Trans.). University of Chicago Press.

Jackson, L.M. (2020). The racial politics of Kamala Harris's performance style. *The New Yorker*. https://www.newyorker.com/culture/annals-of-appearances/the-racial-politics-of-kamala-harriss-performance-style.

Jackson, M. (2005). *Existential anthropology: Events, exigencies, and effects*. Berghahn.

Jackson, M. (2009). *The palm at the end of the mind: Relatedness, religiosity, and the real*. Duke University Press.

Jackson, M. (2013a). *Lifeworlds: Essays in existential anthropology*. University of Chicago Press.

Jackson, M. (2013b). *The other shore: Essays on writers and writing*. University of California Press.

Jackson, P. (1998). Constructions of "whiteness" in the geographical imagination. *Area, 30*(2), 99–106. https://doi.org/10.1111/j.1475-4762.1998.tb00053.x.

James, P.E. (1942). *Latin America*. Cassell.

Jamieson, W. (2021). One or several granular geographies? *Dialogues in Human Geography, 11*(2), 311–14. https://doi.org/10.1177/20438206211004860.
Jazeel, T. (2014). Subaltern geographies: Geographical knowledge and postcolonial strategy. *Singapore Journal of Tropical Geography, 35*(1), 88–103. https://doi.org/10.1111/sjtg.12053.
Jones, R., & Fowler, C. (2007). Where is Wales? Narrating the territories and borders of the Welsh linguistic nation. *Regional Studies, 41*(1), 89–101. https://doi.org/10.1080/00343400600928343.
Joronen, M. (2012). Heidegger on the History of machination. *Critical Horizons, 13*(3), 351–76. https://doi.org/10.1558/crit.v13i3.351.
Joronen, M. (2013). Heidegger, event and the ontological politics of the site. *Transactions of the Institute of British Geographers, 38*(4), 627–38. https://doi.org/10.1111/j.1475-5661.2012.00550.x.
Joronen, M. (2017). Spaces of waiting: Politics of precarious recognition in the occupied West Bank. *Environment and Planning D: Society and Space, 35*(6), 994–1011. https://doi.org/10.1177/0263775817708789.
Joronen, M. (2021a). Playing with plenitude and finitude: Attuning to a mysterious void of being. In A.J. Secor & P. Kingsbury (Eds.), *A place more void* (pp. 249–66). University of Nebraska Press.
Joronen, M. (2021b). To wound life, to prevent recovery: enforcing vulnerability in Gaza. In D. Bissell, M. Rose, & P. Harrison (Eds.), *Negative geographies: Exploring the politics of limits* (pp. 206–33). University of Nebraska Press.
Joronen, M. (2021c). Unspectacular spaces of slow wounding in Palestine. *Transactions of the Institute of British Geographers, 46*(4), 995–1007. https://doi.org/10.1111/tran.12473.
Joronen, M., & Rose, M. (2021). Vulnerability and its politics: Precarity and the woundedness of power. *Progress in Human Geography, 45*(6), 1402–18. https://doi.org/10.1177/0309132520973444.
Katz, C.E. (2003). *Levinas, Judaism, and the feminine: The silent footsteps of Rebecca.* Indiana University Press.
Kim, H. (2019). Decolonization and the ontological turn of sociology. *Journal of Asian Sociology, 48*(4), 443–54. https://doi.org/10.1177/02632764211073011.
Knappett, C. (2005). *Thinking through material culture: An interdisciplinary perspective.* University of Pennsylvania Press.
Knauft, B. (2006). Anthropology in the middle. *Anthropological Theory, 6*(4), 407–30. https://doi.org/10.1177/1463499606071594.
Kohn, E. (2013). *How forests think: Toward an anthropology beyond the human.* University of California Press.
Kong, L. (1999). Cemetaries and columbaria, memorials and mausoleums: Narrative and interpretation in the study of deathscapes in geography. *Australian Geographical Studies, 37*(1), 1–10.
Kroeber, A.L. (1917). The superorganic. *American Anthropologist, 19*(2), 163–213.

Kroeber, A.L. (1952). *The nature of culture*. University of Chicago Press.
Kuper, A. (1999). *Culture: The anthropologists' account*. Harvard University Press.
Lacan, J. (1953). Some reflections on the ego. *International Journal of Psychoanalysis, 34*, 11–17.
Lacan, J. (1988). *The ego in Freud's theory and in the technique of psychoanalysis, 1954–1955*. W.W. Norton.
Lacan, J. (2006). The function and field of speech and language in psychoanalysis (B. Fink, H. Fink, & R. Grigg, Trans.). In *Ecrits: The first complete edition in English*. W.W. Norton & Co.
Lacan, J. (2008). *The ethics of psychoanalysis, 1959–1960: The seminar of Jacques Lacan, book VII*. Routledge.
Lafrance, A. (2017). The normalisation of conspiracy culture. *The Atlantic*. https://www.theatlantic.com/technology/archive/2017/06/the-normalization-of-conspiracy-culture/530688/.
Laidlaw, J. (2012). Ontologically challenged. *Anthropology of This Century, 4*. http://aotcpress.com/articles/ontologically-challenged/.
Lamarck, J.B. (1914). *Zoological philosophy*. Macmillan and Company.
Laplanche, J. (1999). *Essays on otherness* (L. Thurston, Trans. & J. Fletcher, Ed.). Routledge.
Large, W. (2015). *Levinas' "Totality and infinity": A reader's guide*. Bloomsbury.
Larsen, S.C., & Johnson, J.T. (2012). In between worlds: Place, experience, and research in Indigenous geography. *Journal of Cultural Geography, 29*(1), 1–13. https://doi.org/10.1080/08873631.2012.646887.
Latour, B. (1987). *Science in action: How to follow scientists and engineers through society*. Harvard University Press.
Latour, B. (1988). *The pasteurization of France* (A. Sheridan & J. Law, Trans.). Harvard University Press.
Latour, B. (1993). *We have never been modern* (C. Porter, Trans.). Harvard University Press.
Latour, B. (1997). On actor-network theory: A few clarifications. *Soziale Welt*, 47(4), 369–81. http://www.jstor.org/stable/40878163.
Latour, B. (2004). *Politics of nature: How to bring the sciences into democracy* (C. Porter, Trans.). Harvard University Press.
Latour, B. (2013). *An inquiry into modes of existence: An anthropology of the moderns*. Harvard University Press.
Lawson, T. (2003). Ontology and feminist theorizing. *Feminist Economics, 9*(1), 119–50. https://doi.org/10.1080/1354570022000035760.
Levinas, E. (1969). *Totality and infinity*. Duquesne University Press.
Levinas, E. (1981). *Otherwise than being, or beyond essence*. University of Duquesne Press.
Levinas, E. (1987). *Time and the other and additional essays* (R.A. Cohen, Trans.). Duquesne University Press.

Levinas, E. (1996a). Meaning and sense. In A.T. Peperzak, S. Critchley, & R. Bernasconi (Eds.), *Emmanuel Levinas: Basic philosophical writings* (pp. 34–64). University of Indiana Press.

Levinas, E. (1996b). Transcendence and height. In A.T. Peperzak, S. Critchley, & R. Bernasconi (Eds.), *Emmanuel Levinas: Basic philosophical writings* (pp. 11–32). Indiana University Press.

Levinas, E. (1998). *Entre nous: On thinking-of-the-other*. Columbia University Press.

Levinas, E. (2019). The animal interview. In P. Atterton & T. Wright (Eds.), *Face-to-face with animals : Levinas and the animal question* (pp. 21–4). State University of New York.

Levi-Strauss, C. (1969). *The raw and the cooked: Introduction to a science of mythology*. Pimlico.

Lewis, H. (2020). How capitalism drives cancel culture. *The Atlantic*. https://www.theatlantic.com/international/archive/2020/07/cancel-culture-and-problem-woke-capitalism/614086/.

Livingstone, D.N. (1993). *The geographical tradition: Episodes in the history of a contested enterprise*. Blackwell Publishers.

Lorimer, H. (2006). Herding memories of humans and animals. *Environment and Planning D: Society and Space*, *24*(4), 497–518. https://doi.org/10.1068/d381t.

Lorimer, H., & Lund, K. (2008). A collectable topography: Walking, remembering and recording mountains. In T. Ingold & J.L. Vergunst (Eds.), *Ways of walking: Ethnography and practice on foot* (pp. 185–200). Ashgate.

Lorimer, J. (2012). Aesthetics for post-human worlds: Difference, expertise and ethics. *Dialogues in Human Geography*, 2(3), 284–7. https://doi.org/10.1177/2043820612468646.

Lossau, J. (2009). Anthropogeography (after Ratzel). In R. Kitchin & N. Thrift (Eds.), *International encyclopedia of human geography* (pp. 140–7). Elsevier.

Lukianoff, G., & Haidt, J. (2018). *The coddling of the American mind: How good intentions and bad ideas are setting up a generation for failure*. Penguin Press.

Lund, K. (2008). Listen to the sound of time: Walking with saints in an Andalusian village. In T. Ingold & J.L. Vergunst (Eds.), *Ways of walking: Ethnography and practice on foot* (pp. 93–104). Ashgate.

Malabou, C. (2008). *What should we do with our brain?* Fordham University Press.

Malafouris, L. (2004). The cognitive basis of material engagement: where brain, body and culture conflate. In E. DeMarrais, C. Gosden, & C. Renfrew (Eds.), *Rethinking materiality: The engagement of mind with the material world* (pp. 53–62). McDonald Institute for Archaeological Research.

Malafouris, L. (2013). *How things shape the mind: A theory of material engagement*. MIT Press.

Malik, K. (2017). In defense of cultural appropriation. *The New York Times*. https://www.nytimes.com/2017/06/14/opinion/in-defense-of-cultural-appropriation.html?searchResultPosition=2.

Malinowski, B. (1961). *Argonauts of the western Pacific: An account of native enterprise and adventure in the archipelagoes of Melanisian New Guinea*. Dutton.

Malpas, J.E. (2008). Heidegger, geography and politics. *Journal of the Philosophy of History*, *2*, 185–213.

Malpas, J. (2012). *Heidegger and the thinking of place: Explorations in the topology of being*. MIT Press.

Malpas, J.E. (2006). *Heidegger's topology: Being, place, world*. MIT Press.

Manning, J., & Campbell, B. (2014). Microagression and moral cutlures. *Comparative Sociology*, *13*(6), 669–726.

Marcus, G.E. (2008). The end(s) of ethnography: Social/cultural anthropology's signature form of producing knowledge in transition. *Cultural Anthropology*, *23*(1), 1–14. https://doi.org/10.1111/j.1548-1360.2008.00001.x.

Marres, N. (2009). Testing powers of engagement: Green living experiments, the ontological turn and the undoability of involvement. *European Journal of Social Theory*, *12*(1), 117–33. https://doi.org/10.1177/1368431008099647.

Marston, S.A., Jones III, J.P., & Woodward, K. (2005). Human geography without scale. *Transactions of the Institute of British Geographers*, *30*(4), 416–32. https://doi.org/10.1111/j.1475-5661.2005.00180.x.

Mathewson, K. (2011). Sauer's Berkeley School legacy: Foundation for an emergent environmental geography. In G. Bocco, P.S. Urquijo, & A. Vieyra (Eds.), *Geografia Y Ambiente en America Latina* (pp. 51–81). Centro de Investigaciones en Geografia Ambiental.

Matless, D. (1992a). An occasion for geography: Landscape, representation, and Foucault's corpus. *Environment and Planning D: Society and Space*, *10*(1), 41–56. https://doi.org/10.1068/d100041.

Matless, D. (1992b). Regional surveys and local knowledges: The geographical imagination in Britain. 1918–1939. *Transactions Institute of British Geographers*, *17*(4), 464–80. https://doi.org/10.2307/622711.

Matless, D. (1994). Doing the English village, 1945–1990: An essay in imaginative geography. In P. Cloke, M. Dole, D. Matless, M. Phillips, & N. Thrift (Eds.), *Writing the rural: Five cultural geographies* (pp. 7–88). Paul Chapman.

Matless, D. (1998). *Landscape and Englishness*. Reaktion.

Mattingly, C. (1998). *Healing dramas and clinical plots: The narrative structure of experience*. Cambridge University Press.

McClintock, A. (1995). *Imperial leather: Race, gender, and sexuality in the colonial conquest*. Routledge.

McCormack, D. (2003). An event in geographical ethics in spaces of affect. *Transactions for the Institute for British Geographers*, *11*, 238–47. https://doi.org/10.1111/j.0020-2754.2003.00106.x.

McCormack, D. (2007). Molecular affects in human geographies. *Environment and Planning A, 39*, 359–77. https://doi.org/10.1068/a3889.

McFarlane, C. (2009). Translocal assemblages: Space, power and social movements. *Geoforum, 40*(4), 561–7. https://doi.org/10.1016/j.geoforum.2009.05.003.

McGhee, H.C. (2021). *The sum of us: What racism costs everyone and how we can prosper together*. One World.

McNeil, M. (2010). Post-millennial feminist theory: Encounters with humanism, materialism, critique, nature, biology and Darwin. *Journal for Cultural Research, 14*(4), 427–37. https://doi.org/10.1080/14797581003765382.

McNeill, W. (2006). *The time of life: Heidegger and ethos*. State University of New York Press.

McWhorter, J. (2020). Academics are really, really worried about their freedom. In *The Atlantic*. https://www.theatlantic.com/ideas/archive/2020/09/academics-are-really-really-worried-about-their-freedom/615724/.

Mendieta, E. (2017). Metaphysical anti-Semitism and worldlessness: On world poorness, world forming, and world destroying. In A.J. Mitchell & P. Trawny (Eds.), *Heidegger's black notebooks: Responses to anti-Semitism* (pp. 36–51). Columbia University Press.

Mercier, G. (1995). La région et l'État selon Friedrich Ratzel et Paul Vidal de la Blache. *Annales de géographie, 583*, 211–35. English version available http://www.siue.edu/GEOGRAPHY/ONLINE/mercier.htm#N_1_.

Mercier, G. (2009). Vidal de la Blache. In P. Dans, R. Kitchin, & N. Thrift (Eds.), *International encyclopaedia of human geography* (pp. 147–50). Elsevier.

Merleau-Ponty, M. (2002). *Phenomenology of perception*. Routledge.

Merriman, P. (2012). Human geography without time-space. *Transactions of the Institute of British Geographers, 37*(1), 13–27. https://www.jstor.org/stable/41427925.

Miller, A.S. (2013). *Speculative grace: Bruno Latour and object-oriented theology*. Fordham University Press.

Miller, D. (1994). *Material culture and mass consumption*. Blackwell.

Miller, D. (1998). *A theory of shopping*. Polity Press.

Miller, D. (2005). Materiality: An introduction. In D. Miller (Ed.), *Materiality* (pp. 1–50). Duke University Press.

Miller, D. (2008). *The comfort of things*. Polity.

Miller, D., & Tilley, C. (2011). Replies to Mitch Rose: "Secular materialism: A critique of earthly theory." *Journal of Material Culture, 16*(3), 325–32. https://doi.org/10.1177/1359183511413650.

Miller, J.-A. (2019). *Paradigms of jouissance*. London Society of the New Lacanian School.

Minkov, M., & Hofstede, G. (2012). Is national culture a meaningful concept? Cultural values delineate homogeneous national clusters of in country regions. *Cross-Cultural Research, 46*(2), 133–59. https://doi.org/10.1177/1069397111427262.

Mitchell, D. (1995). There's no such thing as culture: Towards a reconceptualization of the idea of culture in geography. *Transactions of the Institute for British Geographers, 20*, 102–16. https://doi.org/10.2307/622727.

Mitchell, D. (2000). *Cultural geography: A critical introduction*. Blackwell.

Mitzen, J. (2006). Ontological security in world politics: State identity and the security dilemma. *European Journal of International Relations, 12*(3), 341–70. https://doi.org/10.1177/1354066106067346.

Morrison, T. (1987). *Beloved: A novel*. Knopf.

Mouffe, C. (2013). *Agonistics: Thinking the world politically*. Verso.

Nancy, J.-L. (2000). *L'Intrus* [The intruder]. *Galilée*.

Natter, W. (2005). Friedrich Ratzel's spatial turn. In H.V. Houtum, O. Kramsch, & W. Zierhofer (Eds.), *B/ordering space* (pp. 179–88). Ashgate.

Noxolo, P. (2017). Introduction: Decolonising geographical knowledge in a colonised and re-colonising postcolonial world. *Area, 49*(3), 317–19. https://doi.org/10.1111/area.12370.

Ogborn, M. (1992). Local power and state regulation in nineteenth-century Britain. *Transactions of the Institute of British Geographers, 17*(2), 215–26. https://doi.org/10.2307/622547.

Ogborn, M. (1995). Discipline, government and law: Separate confinement in the prisons of England and Wales, 1830–1877. *Transactions of the Institute of British Geographers, 20*(3), 295–311. https://doi.org/10.2307/622653.

Ortner, S.B. (1999). *The fate of "culture": Geertz and beyond*. University of California Press.

Oswin, N. (2020). An other geography. *Dialogues in Human Geography, 10*(1), 9–18. https://doi.org/10.1177/2043820619890433.

Paasi, A. (2003). Region and place: Regional identity in question. *Progress in Human Geography, 27*(4), 475–85. https://doi.org/10.1191/0309132503ph439pr.

Panelli, R. (2008). Social geographies: Encounters with Indigenous and more-than-White/Anglo geographies. *Progress in Human Geography, 32*(6), 801–11. https://doi.org/10.1177/0309132507088031.

Peake, L.J. (2017). Anthropogeography. In D. Richardson, N. Castree, M.F. Goodchild, A. Kobayashi, W. Liu, & R.A. Marston (Eds.), *International encyclopedia of geography: People, the earth, environment and technology* (pp. 1–3). Wiley.

Peck, E. (2018). 5 women sue Monster Energy over abusive, discriminatory culture. *Huffington Post*. https://www.huffpost.com/entry/monster-energy-lawsuits_n_5a6280c1e4b002283002ca27.

Pedersen, M.A. (2011). *Not quite shamans: Spirit worlds and political lives in northern Mongolia*. Cornell University Press.

Pedersen, M.A. (2012a). Common nonsense: A review of certain recent reviews of the "ontological turn." *Anthropology of This Century, 5.* http://aotcpress.com/articles/common_nonsense/.

Pedersen, M.A. (2012b). Proposing the motion: Morten Axel Pedersen. *Critique of Anthropology, 32*(1), 59–65. https://doi.org/10.1177/0308275X11430873c.

Pedwell, C., & Whitehead, A. (2012). Affecting feminism: Questions of feeling in feminist theory. *Feminist Theory, 13*(2), 115–29. https://doi.org/10.1177/1464700112442635.

Peperzak, A.T. (1993). *To the other: An introduction to the philosophy of Emmanuel Levinas*. Purdue University Press.

Petri, R. (2016). The Mediterranean metaphor in early geopolitical writings. *History, 101*, 671–91. https://doi.org/10.1111/1468-229X.12326.

Pham, M.-h. T. (2014). Fashion's cultural appropriation debate: Pointless. *The Atlantic*. https://www.theatlantic.com/entertainment/archive/2014/05/cultural-appropriation-in-fashion-stop-talking-about-it/370826/.

Phillips, A. (1993). *On kissing, tickling, and being bored: Psychoanalytic essays on the unexamined life*. Harvard University Press.

Philo, C. (1992). Foucault's geography. *Environment and Planning D: Society and Space, 10*(2), 137–61. https://doi.org/10.1068/d100137.

Philo, C. (2017). Less-than-human geographies. *Political Geography, 60*, 256–8. https://doi.org/10.1016/j.polgeo.2016.11.014.

Philo, C. (2021a). Negative geography: "Everything ... less than the universe is subject to suffering." In D. Bissell, M. Rose, & P. Harrison (Eds.), *Negative geographies: Exploring the politics of limits* (pp. 39–66). University of Nebraska Press.

Philo, C. (2021). Nothing-much geographies, or towards micrological investigations. *Geographische Zeitschrift, 109*(2–3), pp. 73–95. doi: 10.25162/gz-2021-0006.

Pickering, A. (2017). The ontological turn. *Social Analysis, 61*(2), 134–50. https://doi.org/10.3167/sa.2017.610209.

Polt, R. (2013). The secret homeland of speech: Heidegger on language 1933–1934. In J.L. Powell (Ed.), *Heidegger and language* (pp. 63–85). Indiana University Press.

Povinelli, E. A. (2002). *The cunning of recognition: Indigenous alterities and the making of Australian multiculturalism*. Duke University Press.

Powell, J. (Ed.). (2013a). *Heidegger and language*. Indiana University Press.

Powell, J. (2013b). Introduction. In J. Powell (Ed.), *Heidegger and language*. Indiana University Press.

Powell, R.C. (2015). The study of geography? Franz Boas and his canonical turn. *Journal of Historical Geography, 49*, 21–30. https://doi.org/10.1016/j.jhg.2015.04.018.

Pratt, M.L. (1992). *Imperial eyes travel writing and transculturation*. Routledge.

Pugh, J., & Chandler, D. (2023). *The world as abyss: The Caribbean and critical thought in the Anthropocene.* University of Westminster Press.

Radcliffe, S.A. (2017). Geography and Indigeneity I: Indigeneity, coloniality and knowledge. *Progress in Human Geography, 41*(2), 220–9. https://doi.org/10.1177/0309132515612952.

Radu, C. (2010). Beyond border-"dwelling": Temporalizing the border-space through events. *Anthropological Theory, 10*(4), 409–33. https://doi.org/10.1177/1463499610386664.

Ratzel, F. (1896). *The history of mankind* (A.J. Butler, Trans.). Macmillan.

Ratzel, F. (1906). Kleine Schriften von Friedrich Ratzel. In H.F. Helmolt (Ed.), *Ausgewaehlt and herausgegeben durch Helmolt, H. Mit einer Bibliographie von Viktor Hantzsch.* Oldenbourg.

Ratzel, F. (1921). *Anthropogeographie* (4th ed.). J. Engelhorns Nachf.

Ratzel, F., & Oberhummer, E. (1923). *Politische geographie* (3rd ed.). R. Oldenbourg.

Ravaisson, F. (Ed.). (2008). *Of habit.* Continuum.

Relph, E. (1976). *Place and placelessness.* Pion.

Renfrew, C. (2004). Towards a theory of material engagement. In E. DeMarrais, C. Gosden, & C. Renfrew (Eds.), *Rethinking materiality: The engagement of mind with the material world* (pp. 23–32). McDonald Institute for Archaeological Research.

Richardson, L. (1988). The collective story: Postmodernism and the writing of sociology. *Sociological Focus, 21*(3), 199–208. https://doi.org/10.1080/00380237.1988.10570978.

Robbins, P. (2004). Cultural ecology. In J.S. Duncan, N.C. Johnson, & R.H. Schein (Eds.), *A companion to cultural geography* (pp. 180–93). Blackwell.

Roberts, T. (2012). From "new materialism" to "machinic assemblage": Agency and affect in IKEA. *Environment and Planning A: Economy and Space, 44*(10), 2512–29. https://doi.org/10.1068/a44692.

Robinson, T. (1986). *Stones of Aran: Pilgrimage.* Faber and Faber.

Robinson, T. (2009). *Stones of Aran: Labyrinth.* NYRB Classics.

Rose, M. (2002). Landscapes and labyrinths. *Geoforum, 33*(4), 445–67. https://doi.org/10.1016/S0016-7185(02)00030-1.

Rose, M. (2011a). A reply to Daniel Miller and Christopher Tilley. *Journal of Material Culture, 16*(3), 333–5. https://doi.org/10.1177/1359183511415005.

Rose, M. (2011b). Secular materialism: A critique of earthly theory. *Journal of Material Culture, 16*(2), 109–29. https://doi.org/10.1177/1359183511401496.

Rose, M. (2014). Negative governance: Vulnerability, biopolitics and the origins of government. *Transactions of the Institute of British Geographers, 39*(2), 209–23. https://doi.org/10.1111/tran.12028.

Rose, M. (2019). Hesitant democracy: Equality, inequality and the time of politics. *Political Geography, 68*, 101–9. https://doi.org/10.1016/j.polgeo.2018.10.003.

Rose, M. (2021a). The question of culture in cultural geography: Latent legacies and potential futures. *Progress in Human Geography*, *45*(5), 951–71. https://doi.org/10.1177/0309132520950464.

Rose, M. (2021b). Tragic democracy: The politics of submitting to others. In D. Bissell, M. Rose, & P. Harrison (Eds.), *Negative geographies: Exploring the politics of limits* (pp. 261–86). University of Nebraska Press.

Rose, M., Bissell, D., & Harrison, P. (2021). Negative geographies. In D. Bissell, M. Rose, & P. Harrison (Eds.), *Negative geographies: Exploring the politics of limits* (pp. 1–38). University of Nebraska Press.

Ross, L. (2019). I'm a black feminist. I think call-out culture is toxic. *The New York Times*. https://www.nytimes.com/2019/08/17/opinion/sunday/cancel-culture-call-out.html?searchResultPosition=17.

Roth, M. (1996). *The poetics of resistance: Heidegger's line*. Northwestern University Press.

Roth, R. (2009). The challenges of mapping complex indigenous spatiality: from abstract space to dwelling space. *Cultural Geographies*, *16*(2), 207–27. https://doi.org/10.1177/1474474008101517.

Ruti, M. (2015). *Between Levinas and Lacan self, other, ethics*. Bloomsbury Publishing.

Sahlins, M. (1985). *Islands of history*. University of Chicago Press.

Sahlins, M. (1999). Two or three things that I know about culture. *Journal of the Royal Anthropological Institute*, *5*(3), 399–421. https://doi.org/10.2307/2661275.

Said, E.W. (1979). *Orientalism*. Vintage Books.

Sandel, M.J. (2020). *The tyranny of merit: What's become of the common good?* Farrar, Straus and Giroux.

Sauer, C.O. (1925). *The morphology of landscape*. University of California Press.

Sauer, C.O. (1941). The personality of Mexico. *Geographical Review*, *31*(3), 353–64.

Saul, S. (2017). "'Victim feminism" and sexaual assult on campus. *The New York Times*. https://www.nytimes.com/2017/11/03/education/edlife/christina-hoff-sommers-sexual-assault-feminism.html?searchResultPosition=4.

Savransky, M. (2017). A decolonial imagination: Sociology, anthropology and the politics of reality. *Sociology*, *51*(1), 11–26. https://doi.org/10.1177/0038038516656983.

Schein, R.H. (1997). The place of landscape: a conceptual framework for interpreting an American scene. *Annals of the Association of American Geographers*, *87*(4), 660–80. https://doi.org/10.1111/1467-8306.00072.

Schor, N. (1989). This essentialism which is not one: Coming to grips with Irigaray. *differences*, *1*(2), 38–58. https://doi.org/10.1215/10407391-1-2-38.

Scott, M. (2007). *The severed snake: Matrilineages, making place, and a Melanesian Christianity in southeast Solomon Islands*. North Carolina University Press.

Secor, A.J., & Kingsbury, P. (2021). Introduction: Into the void. In A.J. Secor & P. Kingsbury (Eds.), *A place more void* (pp. 1–26). University of Nebraska Press.

Sehgal, P. (2015). Is cultural appropriation always wrong? *The New York Times Magazine*. https://www.nytimes.com/2015/10/04/magazine/is-cultural-appropriation-always-wrong.html?searchResultPosition=6.

Semple, E.C. (1968). *Influences of geographic environment, on the basis of Ratzel's system of anthropo-geography*. Russell & Russell.

Serres, M., & Latour, B. (1995). *Conversations on science, culture, and time*. University of Michigan Press.

Sewell, W.H., Jr. (2005). The concept(s) of culture. In G.M. Spiegel (Ed.), *Practicing history: New directions in historical writing after the linguistic turn* (pp. 76–95). Routledge.

Shapiro, K. (2009). Reviving habit: Ravaisson's practical metaphysics. *Theory and Event*, *12*(4). https://doi.org/10.1353/tae.0.0100.

Sharp, J.P. (1996). Hegemony, popular culture and geopolitics: The Reader's Digest and the construction of danger. *Political Geography*, *15*(6), 557–70. https://doi.org/10.1016/0962-6298(96)00031-5.

Sheehan, T. (1981). On movement and the destruction of ontology. *The Monist*, *64*, 534–42. https://doi.org/10.5840/monist198164434.

Sismondo, S. (2015). Ontological turns, turnoffs and roundabouts. *Social Studies of Science*, *45*(3), 441–8. https://doi.org/10.1177/0306312715574681. Medline:26477200.

Sivado, A. (2015). The shape of things to come? Reflections on the ontological turn in anthropology. *Philosophy of the Social Sciences*, *45*(1), 83–99. https://doi.org/10.1177/0048393114524830.

Smith, P.B. (2004). Nations, cultures, and individuals: new perspectives and old dilemmas. *Journal of Cross-Cultural Psychology*, *35*(1), 6–12. https://doi.org/10.1177/0022022103260460

Smith, W.D. (1980). Friedrich Ratzel and the origins of Lebansraum. *German Studies Review*, *3*(1), 51–68. https://doi.org/10.2307/1429483.

Soja, E.W. (1988). *Postmodern geographies: The reassertion of space in critical social theory*. Verso.

Soja, E.W. (1996). *Thirdspace: Journeys to Los Angeles and other real-and-imagined places*. Blackwell.

Sommers, C.H. (2000). *The war against boys: How misguided feminism is harming our young men*. Simon and Schuster.

Speth, W.W. (1978). The anthropogeographic theory of Franz Boas. *Anthropos*, *73*(1/2), 1–31. https://www.jstor.org/stable/40459221.

Spiegel, G.M. (2005). *Practicing history: New directions in historical writing after the linguistic turn*. Routledge.

Spivak, G.C. (1996). Subaltern studies: deconstructing historiagraphy. In D. Landry & G. MacLean (Eds.), *The Spivak reader* (pp. 203–35). Routledge.

Stalans, L., & Finn, M. (2018). Pimps hide in plain sight in corporate America's boys club. *Huffington Post*. https://www.huffpost.com/entry/opinion-stalans-finn-corporate-sex-trade_n_5aac3360e4b05b2217fec64c.

Stengers, I. (2010). *Cosmopolitics*. University of Minnesota Press.
Stewart, K. (1996). *A space on the side of the road: Cultural poetics in an "other" America*. Princeton University Press.
Stewart, K. (2007). *Ordinary affects*. Duke University Press.
Stiegler, B. (1998). *Technics and time*. Stanford University Press.
Stoller, P., & Olkes, C. (1987). *In sorcery's shadow: A memoir of apprenticeship among the Songhay of Niger*. University of Chicago Press.
Strathern, M. (1991). *Partial connections*. Rowman & Littlefield Publishers.
Strathern, M. (2018). Opening up relations. In M. de la Cadena & M. Blaser (Eds.), *A world of many worlds* (pp. 23–52). Duke University Press.
Strohmayer, U. (2021). Urban renewal and the actuality of absence. The "Hole" (trou) of Paris, 1973. In P. Kingsbury & A.J. Secor (Eds.), *A place more void* (pp. 29–47). University of Nebraska Press.
Stroope, S., Froese, P., Rackin, H. M., & Delehanty, J. (2021). Unchurched Christian nationalism and the 2016 U.S. presidential election. *Sociological Forum*, *36*(2), 405–25. https://doi.org/10.1111/socf.12684.
Subotić, J. (2016). Narrative, ontological security, and foreign policy change 1. *Foreign Policy Analysis*, *12*(4), 610–27. https://doi.org/10.1111/fpa.12089.
Sundberg, J. (2014). Decolonizing posthumanist geographies. *Cultural Geographies*, *21*(1), 33–47. https://doi.org/10.1177/1474474013486067.
Suzman, J. (2020). *Work: A history of how we spend our time*. London, Bloomsbury Circus.
Talayesva, D.C. (1963). *Sun Chief: The autobiography of a Hopi Indian*. Yale University Press.
Taussig, M.T. (2004). *My cocaine museum*. University of Chicago Press.
Thrift, N. (2008). *Non-representational theory: Space, politics, affect*. Routledge.
Thrift, N., & Dewsbury, J.-D. (2000). Dead geographies – and how to make them live. *Environment and Planning D: Society and Space*, *18*, 411–32. https://doi.org/10.1068/d1804ed.
Thrift, N., & Whatmore, S. (2004). *Cultural geography: Critical concepts in the social sciences*. Routledge.
Tilley, C. (2004). *The materiality of stone: Explorations in landscape phenomenology*. Berg.
Traldi, O. (2020). On the limits of Dave Rubin's cultural politics. *National Review*. https://www.nationalreview.com/2020/06/book-review-dont-burn-this-book-dave-rubin/.
Trouillot, M.-R. (2003). *Global transformations: Anthropology and the modern world*. Palgrave Macmillan.
Tsing, A.L. (2015). *The mushroom at the end of the world: On the possibility of life in capitalist ruins*. Princeton University Press.
Tuan, Y.-F. (1979). Thought and landscape: The eye and the mind's eye. In D. W. Meinig (Ed.), *The interpretation of ordinary landscapes* (pp. 89–102). Oxford University Press.

Turnock, D. (1967). The region in modern geography. *Geography, 52*(4), 374–83.

Uscinski, J.E., & Parent, J.M. (2014). *American conspiracy theories*. Oxford University Press.

Van Heur, B., Leydesdorff, L., & Wyatt, S. (2013). Turning to ontology in STS? Turning to STS through "ontology." *Social Studies of Science, 43*(3), 341–62. https://doi.org/10.1177/0306312712458144.

Vemuri, A. (2018). "Calling out" campus sexual violence: Student activist labors of confrontation and care. *Communication, Culture and Critique, 11*(3), 498–502. https://doi.org/10.1093/ccc/tcy021.

Verne, J. (2017). The neglected "gift" of Ratzel for/from the Indian Ocean: Thoughts on mobilities, materialities and relational spaces. *Geographica Helvetica, 72*, 85–92.https://doi.org/10.5194/gh-72-85-2017, 2017.

Vidal de la Blache, P. (1902). Les conditions geographiques des faits sociaux. *Annales de géographie, 55*, 13–23.

Vidal de la Blache, P. (1911). Les gendre de vie dans la geographie humaine. *Annales de géographie, 112*, 289–304.

Vidal de la Blache, P. (1926). *Principles of human geography*. Constable.

Viveiros de Castro, E. (1998). Cosmological deixis and Amerindian perspectivism. *The Journal of the Royal Anthropological Institute of Great Britain and Ireland, 4*(3), 469–88. https://doi.org/10.2307/3034157.

Viveiros de Castro, E. (2003). And: After-dinner speech given at Anthropology and Science, the 5th Decennial Conference of the Association of Social Anthropologists of the UK and Commonwealth, 2003. *Issue 7 of Manchester papers in social anthropology*. University of Manchester.

Viveiros de Castro, E. (2004a). Exchanging perspectives: The transformation of objects into subjects in Amerindian ontologies. *Common Knowledge, 10*(3), 463–84. https://doi.org/10.1215/0961754X-10-3-463.

Viveiros de Castro, E. (2004b). Perspectival anthropology and the method of controlled equivocation. *Tipiti: Journal of the Society for the Anthropology of Lowland South America, 2*(1), 3–22.

Viveiros de Castro, E. (2012). *Cosmological perspectivism in Amazonia and elsewhere: four Lectures given in the Department of Social Anthropology, University of Cambridge, February-March 1998*. HAU.

Viveiros de Castro, E. (2014). *Cannibal metaphysics: For a post-structural anthropology*. Univocal.

Viveiros de Castro, E. (2015). Who is afraid of the ontological wolf? Some comments on an ongoing anthropological debate. *The Cambridge Journal of Anthropology, 33*(1), 2–17. https://doi.org/10.3167/ca.2015.330102.

Vollebergh, A. (2016). The other neighbour paradox: Fantasies and frustrations of "living together" in Antwerp. *Patterns of Prejudice, 50*(2), 129–49. https://doi.org/10.1080/0031322X.2016.1161957.

Von Uexküll, J.J. (1957). A stroll through the worlds of animals and men. In C.H. Schiller (Ed.), *Instinctive behavior: The development of a modern concept* (pp. 5–80). International Universities Press.

Wagner, P.L., & Mikesell, M. (Eds.). (1962). *Readings in cultural geography*. University of Chicago Press.

Wagner, R. (1981). *The invention of culture* (Rev. and expanded ed.). University of Chicago Press.

Wagner, R. (2001). *An anthropology of the subject: A holographic worldview in New Guinea and its meaning and significance for the world of anthropology*. University of California Press.

Waldman, K. (2019). In Y.A., where is the line between criticism and cancel culture? *The New Yorker*. https://www.newyorker.com/books/under-review/in-ya-where-is-the-line-between-criticism-and-cancel-culture.

Washburn, M. (2020). After Harvey Weinstein, will we see a cultural change? *National Review*. https://www.nationalreview.com/2020/06/harvey-weinstein-metoo-movement-seeks-cultural-change/.

Weiner, G. (2020). Cancel culture is not the problem; conformity culture is. *National Review*. https://www.nationalreview.com/2020/09/cancel-culture-conformity-culture-bigger-problem/.

Whatmore, S. (2002). *Hybrid geographies: Natures cultures spaces*. Sage.

Whatmore, S. (2006). Materialist returns: Practicing cultural geography in and for a more-than-representational world. *Cultural Geographies*, *13*(4), 600–9. https://doi.org/10.1191/1474474006cgj377oa.

Whitehead, A.L., & Perry, S.L. (2020). *Taking America back for God: Christian nationalism in the United States*. Oxford University Press.

Whitehead, A.L., Perry, S.L., & Baker, J.O. (2018). Make America Christian again: Christian nationalism and voting for Donald Trump in the 2016 presidential election. *Sociology of Religion*, *79*(2), 147–71. https://doi.org/10.1093/socrel/srx070.

Will, G.F. (2017). The left's misguided obsession with "cultural appropriation." *The Washington Post*. https://www.washingtonpost.com/opinions/the-lefts-misguided-obsession-with-cultural-appropriation/2017/05/12/59e518bc-3672-11e7-b4ee-434b6d506b37_story.html.

Williams, R. (1971). *Culture and society: 1780–1950*. Penguin.

Williams, R. (1977). *Marxism and literature*. Oxford University Press.

Williams, R. (1981). *Culture*. Fontana Press.

Williams, R. (1995). *The sociology of culture*. University of Chicago Press.

Williams, R. (2003). The analysis of culture. In C. Jenks (Ed.), *Culture: Critical concepts in sociology* (pp. 28–50). Routledge.

Williamson, K.D. (2020). Where culture meets money. *National Review*. https://www.nationalreview.com/2020/05/economic-development-wall-street-where-culture-meets-money/.

Wine, N. (2018). The object from Freud to Lacan. In A. Gessert (Ed.), *Introductory lectures on Lacan* (pp. 39–54). Karnac.
Witz, A. (2000). Whose body matters? Feminist sociology and the corporeal turn in sociology and feminism. *Body & Society*, *6*(2), 1–24. https://doi.org/10.1177/1357034X00006002001.
Wolin, R. (1992). *The politics of being*. Columbia University Press.
Wolin, R. (Ed.). (1993). *The Heidegger controversy: A critical reader*. MIT Press.
Woolgar, S., & Lezaun, J. (2015). Missing the (question) mark? What is a turn to ontology? *Social Studies of Science*, *45*(3), 462–7. https://doi.org/10.1177/0306312715584010. Medline:26477203.
Wright, S. (2015). More-than-human, emergent belongings: A weak theory approach. *Progress in Human Geography*, *39*(4), 391–411. https://doi.org/10.1177/0309132514537132.
Wylie, J. (2002). The heart of the visible: An essay on ascending Glastonbury Tor. *Geoforum*, *33*(4), 441–54. https://doi.org/10.1016/S0016-7185(02)00033-7.
Wylie, J. (2005). A single day's walking: Narrating self and landscape on the South West Coast Path. *Transactions for the Institute for British Geographers*, *30*(3), 234–47. https://doi.org/10.1111/j.1475-5661.2005.00163.x.
Wylie, J. (2007). *Landscape*. Routledge.
Wylie, J. (2009). Landscape, absence and the geographies of love. *Transactions of the Institute of British Geographers*, *34*(3), 275–89. https://doi.org/10.1111/j.1475-5661.2009.00351.x.
Wylie, J. (2010). Cultural geographies of the future, or looking rosy and feeling blue. *Cultural Geographies*, *17*(2), 211–17. https://doi.org/10.1177/1474474010363852.
Wylie, J. (2012). Dwelling and displacement: Tim Robinson and the questions of landscape. *Cultural Geographies*, *19*(3), 365–83. https://doi.org/10.1177/1474474011429406.
Wylie, J. (2016a). A landscape cannot be a homeland. *Landscape Research*, *41*(4), 408–16. https://doi.org/10.1080/01426397.2016.1156067.
Wylie, J. (2016b). Timely geographies: "New directions in cultural geography" revisited. *Area*, *48*(3), 374–7. https://doi.org/10.1111/area.12289.
Yar, S., & Bromwich, J.E. (2019). Tales from the teenage cancel culture. *The New York Times*. https://www.nytimes.com/2019/10/31/style/cancel-culture.html?searchResultPosition=1.
Young, J. (2000). What is dwelling? In H.L. Dreyfus, M.A. Wrathall, & J.E. Malpas (Eds.), *Essays in honor of Hubert L. Dreyfus* (pp. 187–203). MIT Press.
Young, J. (2006). The fourfold. In C.B. Guignon (Ed.), *The Cambridge companion to Heidegger* (2nd ed., pp. 373–93). Cambridge University Press.
Yusoff, K. (2015). Geologic subjects: Nonhuman origins, geomorphic aesthetics and the art of becoming inhuman. *Cultural Geographies*, *22*(3), 383–407. https://doi.org/10.1177/1474474014545301.

Zhang, V. (2020). NOISY FIELD EXPOSURES, or what comes before attunement. *Cultural Geographies, 27*(4), 647–64. https://doi.org/10.1177/1474474020909494.

Zhang, V. (2021). Ethics for the unaffirmable: The hesitant love of a cultural translator. In D. Bissell, M. Rose, & P. Harrison (Eds.), *Negative geographies: Exploring the politics of limits* (pp. 92–118). University of Nebraska Press.

Zhang, V. (2023). Desire's misrecognitions, or the promise of mutable attachments. *Dialogues in Human Geography, 13*(3), 410–13. https://doi.org/10.1177/20438206231151424.

Ziarek, K. (1999). Art, power and politics: Heidegger on Machenschaft and Poiēsis. *Heidegger Circle Proceedings, 33*, 52–71. https://doi.org/10.5840/heideggercircle1999333.

Zigon, J. (2014). An ethics of dwelling and a politics of world-building: A critical response to ordinary ethics. *Journal of the Royal Anthropological Institute, 20*(4), 746–64. https://doi.org/10.1111/1467-9655.12133.

Zimmerer, K S. (1996). *Changing fortunes: Biodiversity and peasant livelihood in the Peruvian Andes*. University of California Press.

Zimmerman, M.E. (1986). *Eclipse of the self: The development of Heidegger's concept of authenticity*. Ohio University Press.

Žižek, S. (2002). *For they know not what they do: Enjoyment as a political factor*. Verso.

Žižek, S. (2005). Neighbors and other monsters: A plea for ethical violence. In S. Žižek, E.L. Santner, & K. Reinhard (Eds.), *The neighbor: Three inquiries in political theology* (pp. 134–90). University of Chicago Press.

Žižek, S. (2006). *How to read Lacan*. Granta.

Žižek, S. (2008a). *The sublime object of ideology*. Verso.

Žižek, S. (2008b). Tolerance as an Ideological Category. *Critical Inquiry, 34*(4), 660–82. https://doi.org/10.1086/592539.

Zornberg, A.G. (1995). *Genesis: The beginning of desire*. The Jewish Publication Society.

Zupančič, A. (2017). *What is sex?* MIT Press.

Index